# *Contents*

90 0541027 8

# Analyzing and Solving CHEMISTRY PROBLEMS

## James E. Brady
*St. John's University (Emeritus), New York*

## Frederick Senese
*Frostburg State University, Maryland*

 JOHN WILEY & SONS, INC.

| | |
|---|---|
| ACQUISITIONS EDITOR | Deborah Brennan |
| MARKETING MANAGER | Robert Smith |
| PRODUCTION EDITOR | Barbara Russiello |
| DESIGNER | Dawn L. Stanley |
| PHOTO EDITOR | Hilary Newman |
| COVER ART | Norm Christensen |
| COVER PHOTOS | *Welder:* ©Lynn Johnson/AURORA |
| | *Fireworks:* ©Tony Wiles/Tony Stone Images/New York, Inc. |
| | *Hot-air balloon:* ©Donovan Reese/ Tony Stone Images/New York, Inc. |
| | *Lighthouse:* ©John Lund// Tony Stone Images/New York, Inc. |
| | *Diamond:* ©Color Day Productions/The Image Bank |

This book was set in 11/13 Times Ten by UG / GGS Information Services, Inc. and printed and bound by Victor Graphics, Inc. The cover was printed by Victor Graphics, Inc.

This book is printed on acid free paper. ∞

ISBN 0-471-25496-7 ✓

Printed in the United States of America

10 9 8 7 6 5 4 3 2 1

# INTRODUCTION

If you're like many students studying chemistry, one of the most frustrating obstacles you face is solving chemistry problems. (That's probably why you've turned to this Supplement.) Our goal here is two-fold. First, of course, we want to help you apply chemical concepts to solve chemistry problems. Second, we want to help you develop better problem solving skills in general—skills that you can apply in other settings besides chemistry.

Because of the limited space available here, we will not attempt to cover all of the kinds of chemistry problems you will encounter. Instead, we will focus on three of the most important areas that you're likely to struggle with early in the course—areas that, once mastered, will enable you to easily conquer the rest. These are: 1. Unit conversions and the factor-label method, 2. SI units and conversions, and 3. The mole concept and stoichiometric calculations. We will discuss them in the order in which they are presented in the textbook, so if you encounter something unfamiliar while working with the Supplement, it's probably because you haven't encountered it yet in class or in your reading of the text. We urge you to postpone studying a topic in this Supplement until it's been presented to you in class and until you've studied the topic in the textbook.

## Becoming a Problem Solver

Becoming a good problem solver involves developing a way of *thinking* about problems. You have to look for ways to break the problem down into simpler parts and you have to search your knowledge base for information that you can apply to the specific problem at hand. Sometimes it helps to rephrase the problem in your own words. There are times when you might recognize that there is a critical step that will be necessary at some central point in the problem. If so, then you can look for ways to work both *backwards*, toward the information you have at the start of the problem, and *forwards* toward the answer you seek. One of the most important things to understand is that there is no single, "one way" so solve all problems. You must work on developing your abilities to think, and toward this end, we will provide you with a framework within which to sharpen these abilities.

## How to Use the Supplement

Before we begin, it is imperative that you realize that this Supplement is *not* a replacement for your lecture class or textbook. There are no short-cuts. To earn a good grade in your course you must begin with a positive attitude, and you must attend class. You must also study the textbook, which has been carefully written to assist you in developing the skills you need to learn the subject. You should turn to this Supplement when you feel you need a further in-depth discussion of problem solving.

The Supplement is divided into sections, each dealing with a particular topic. Within each section you will find a Skills Analysis, a brief review of the subject, one or more worked Problems, and a set of Additional Problems.

### Skills Analyses

We urge you to begin your study of a topic by taking the Skills Analysis test. Its function is to identify areas of weakness and suggest how you can better focus your study.

### Worked Problems

The examples provided in this supplement offer more detail than those in the textbook. We will discuss in detail the thinking processes involved in analyzing problems and, where possible, show you how we can break them down into more manageable pieces. As you study these examples, observe how the specific chemical "Tools" are selected and assembled to provide the solutions. Also study how to eliminate errors by checking your answer to determine whether it is reasonable.

### Additional Problems

To provide you with an opportunity to practice what you've learned, we include a set of additional problems arranged in order of increasing difficulty. The answers to these problems are included so you can check your work.

### The "Toolbox" Concept in Problem Solving

Before we begin to discuss specific chemical concepts and problems, let's review the "Toolbox" concept, which is used in problem solving in your textbook. We believe you will find it to be a powerful aid in helping you analyze, understand, and solve problems of all kinds.

In general, solving a problem involves the use of various pieces of knowledge, which are brought to bear on the problem to obtain the solution. For example, to repair a car an auto mechanic uses knowledge about how to apply various tools such as wrenches, screwdrivers, and pliers along with a knowledge of how an auto works. In each step of a repair, the mechanic selects a tool according to what the tool is capable of doing. To loosen a nut, a wrench is used because that's what wrenches are for. To turn a screw, a screwdriver is used because that's the function of a screwdriver. Thus, to solve mechanical problems with a car, a mechanic relies on a knowledge of *how* tools work and the *specific functions* for which they are designed. *Once equipped with such a set of tools, many different problems (repairs) can be tackled.*

Each discipline has a unique set of tools that can be applied to a large variety of problems. To become a proficient problem solver in a particular area, one learns what tools are available, how to use them, and the tasks the tools are designed to accomplish.

And so it is in chemistry. The tools we use to solve chemistry problems, however, are generally not physical tools (like wrenches and screwdrivers). Instead, the tools are usually derived from chemical concepts. Therefore, to help you learn to solve chemistry problems effectively, we will teach you how the tools of chemistry work and help you develop the ability to recognize when particular tools are required in the solution of a problem. The knowledge so acquired becomes your chemistry "toolbox" which you will be able to use to solve many different chemistry problems.

## Problem Analysis and Critical Links

Just owning a set of tools doesn't make you an auto mechanic. To be one, you need to know how a car works in great detail, and acquiring this knowledge requires study and experience. As you gain experience, you learn to recognize certain symptoms that suggest which part in the car isn't working correctly and needs to be replaced. In other words, rather than just taking the car apart, you spend some time analyzing the problem. Once you have identified the malfunctioning part, repairs can begin. We might think of this malfunctioning part as the **critical link** in coming up with a solution to the repair problem. Once the critical link is identified, the path to solving the problem can be planned and the appropriate tools can be selected to perform the job.

In solving chemistry problems, you must resist the temptation to just start "punching numbers" into your calculator (if it's a numerical problem). Like the mechanic, you have to begin by analyzing the problem. Often, success involves finding and recognizing the necessary critical link required to obtain the solution. When you do this, you will frequently find that the problem breaks down into smaller chunks, each of which is easier to tackle. This is especially important in problems that require multiple steps, where the path to the answer isn't obvious. Because this is such a powerful problem solving technique, one of our goals in developing the worked examples is to show you how to find critical links in problems.

## An Example Illustrating the Critical Link Concept

Let's take a look at an example that illustrates how we can use problem solving techniques in a situation that might arise in our daily lives. Consider the following:

> During a cross country trip on Interstate I-80, a driver became concerned that her speedometer was inaccurate. To avoid a speeding ticket she decided to check the speedometer. She adjusted the speed of her car until the speedometer read 65 mph and set the car on speed control so the speed would remain constant. She then used the stopwatch feature of her wristwatch to time how long it took to travel exactly 5 miles, using the mile markers along the side of the road to measure her progress. She found that it took 4 minutes and 24 seconds to travel the 5 miles. She then pulled into a rest stop to calculate her actual speed over the 5-mile stretch. What was her actual speed in miles per hour (mph)?

The problem here is to use the information given to find the driver's speed *expressed in miles per hour*. The critical link in this problem is to realize that the *per* in *miles per hour* means "divided by," so we can calculate the speed by *dividing* the distance traveled, in miles, by the elapsed time, in hours. For example, if she had traveled 60 miles in 2 hours, her speed would be 30 mile per hour (where we have used the singular form of each unit[1]). A common way of writing this, of course, is 30 mph.

$$\frac{60 \text{ mile}}{2 \text{ hour}} = 30 \frac{\text{mile}}{\text{hour}} = 30 \text{ mile per hour} = 30 \text{ mph}$$

In the problem at hand, the distance traveled[2] is 5.00 miles. To obtain the driver's speed, we need to divide this number by the time of travel expressed in hours.

$$\frac{5.00 \text{ mile}}{? \text{ hour}} = \text{speed in miles per hour}$$

The time of travel was 4 minutes and 24 seconds, which we must convert to hours. There's a couple of ways we could do this. We could find what fraction of a minute 24 seconds is, add that to the 4 minutes, and then change the sum to a time in hours. Another way is to change the 4 minutes to seconds, add them to the 24 seconds to get the total time in seconds. Then divide by the number of seconds in an hour to find how many hours. Let's follow this second path.

Since there are 60 seconds in a minute, 4 minutes must be $(4 \times 60)$ seconds = 240 seconds. Adding the 24 seconds to obtain the total time gives 264 seconds. In one hour there are 3600 seconds, so the time in hours equals (264/3600) hour = 0.0733 hour. Now we can calculate the drivers speed:

$$\frac{5.00 \text{ mile}}{0.0733 \text{ hour}} = 68.2 \text{ mile per hour} = 68.2 \text{ mph}$$

According to our answer, the driver's speedometer reads a little low (65 mph when her actual speed is 68.2 mph). However, aside from checking the arithmetic, can we feel confident that we've solved the problem correctly? We can if the answer seems reasonable, and here it does. We don't expect that the speedometer will be grossly in error and therefore we expect that the answer should be fairly close to 65 mph, which it is.

This check for reasonableness should be part of any problem solving you do, because it gives you confidence in your answer and, if you've made an error, may well signal your mistake. For example, if we had made a mistake in setting up the preceding calculation, we might have obtained an answer such as 750 mph. Clearly, that's an absurd speed for a car, and you would know that you had better check the calculation to find your error.

---

[1]When using units in calculations, we generally do not make the distinction between singular and plural. Usually, in scientific calculations the singular form is used.

[2]Notice that we've expressed the distance to three significant figures. We expect the uncertainty in the distance to be less than ±0.1 miles (which would be about ±600 ft), and probably closer to ± 0.01 miles (approximately ±60 ft), so two decimal places in the distance seems warranted.

# 1. UNIT CONVERSIONS AND THE FACTOR-LABEL METHOD

**Skills Analysis**

1. If you can drive a car 150 miles on 5 gallons of gasoline, how many gallons of gasoline will you need to drive 240 miles?

2. Nurses and paramedics control the volume of liquid delivered by an intravenous unit (I.V.) by setting the drip rate. If a 20 mL dose of liquid is to be administered over a 1-hour period, what should the drip rate be set to, in drops per second? (60 drops are equivalent to one milliliter on this particular I.V.)

3. Suppose you have a problem that says "1.00 g of water at 4 °C has a volume of 1.00 mL." Can you write 1.00 g water/4 °C as a valid conversion factor?

4. Write a conversion factor that converts nanometers (nm) to inches. (1 nm is $10^{-9}$ m; 1 cm is $10^{-2}$ m; and 1 in. is 2.54 cm.)

5. If an asteroid is hurtling towards the earth at a speed of 36,000 miles per hour, and the asteroid is just beyond the orbit of the moon (385,000 km), how long will it take the asteroid to collide with the earth? (1 km = 1000 m; 1 m = 39.37 in.; 12 in. = 1 ft; 1 mi = 5280 ft)

6. The number of atoms in a sample of an element is directly proportional to the mass of the sample. If 4 g of He contains $6 \times 10^{23}$ atoms, how many atoms are in 2 g of He?

**Answers**

If you answered all six questions correctly, you can safely skip this section and go on to take the skills analysis for "SI Prefixes." However, if you missed any of the questions you should study the entire section.

**1.** 8 gallons, **2.** 1/3 drop per second, **3.** No. There's no direct proportionality between mass and temperature. **4.** $\dfrac{3.94 \times 10^{-4}\ \text{in.}}{1\ \text{nm}}$, **5.** 6.7 hr, **6.** $3 \times 10^{23}$ atoms

---

**Why Units Are Essential in Scientific Calculations**

Units give meaning to numbers. Without them, numbers make no sense. For example, what have you learned if a friend tells you he has *three*? Three what? Shoes, dollars, French fries? *Units identify the quantity being described.* For measured quantities, *units also provide a sense of scale.* An object 3 meters long is a lot larger than one that is 3 centimeters long. Thus, units serve two purposes, and without them numbers are really meaningless. For this reason, you should always be sure to include the units when you write down a number; for clarity and to follow standard practice, also remember to leave a space between the number and the unit.

3 meters (correct)          3meters (incorrect)

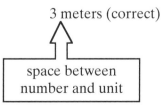

space between
number and unit

For economy of space, units are usually represented by agreed-upon abbreviations or symbols. The meter, for example, is represented by m and centimeter by cm. It's important to use standard abbreviations for units, otherwise someone looking at the number won't be able to understand what it means. We'll review SI units and their abbreviations shortly.

## Using Units to Help Set Up the Solution to a Problem

One of the useful things about units is that they sometimes suggest the solutions to problems. For example, in the example discussed previously in which the driver was checking her speedometer, speed was calculated by forming a fraction with the units *mile* in the numerator and *hour* in the denominator. The units of the desired answer, mile per hour, tell us that we *must* place the number associated with the unit mile in the numerator and the number with the unit hour in the denominator.

$$\frac{5.00 \text{ mile}}{0.0733 \text{ hour}} = 68.2 \text{ mile per hour}$$

In other words, *the units tell us how to set up the arithmetic to obtain the correct answer.*

Let's look at another example. Suppose you wished to calculate the fuel economy of your car in units of *miles per gallon* (or *mile per gallon*, using the singular form of each unit) and that you've determined that your car has consumed 12.6 gallons of fuel while traveling 274 miles. As we've already noted, "per" means "divided by," so in terms of units, mile per gallon means

$$\text{mile} \div \text{gallon} \quad \text{or} \quad \frac{\text{mile}}{\text{gallon}}$$

This tells us that we have to set up a fraction with the units mile in the numerator and gallon in the denominator. In other words, we must divide 274 mile by 12.6 gallon.

$$\frac{274 \text{ mile}}{12.6 \text{ gallon}} = 21.7 \frac{\text{mile}}{\text{gallon}} = 21.7 \text{ mile per gallon}$$

We see here that knowing the correct units of the answer can sometimes direct us toward the solution to the problem.

## Is the Answer Reasonable?

Before we go on, let's once again check our answer to see if it's reasonable. First, for a car, a fuel economy of about 22 miles per gallon (22 mpg) seems reasonable just from common experience. However, in chemistry problems it's likely that you won't have had enough experience to make such qualitative judgments, so let's look at how we can do some approximate arithmetic to see if the answer is in the right "ballpark."

The answer we obtained (21.7 mpg) is not far from 20 mpg, a nice round number we can use to do some simple arithmetic. If the fuel economy were 20 miles per gallon, then we could travel 20 miles on 1 gallon of fuel. Using 12.6 gallons, we could travel a distance of (12.6 × 20) miles, or 252 miles[3]. The actual distance traveled was 274 miles, so our

---

[3]This is the kind of arithmetic you can learn to do in your head. Consider this: if the fuel economy were 10 mpg, then 12.6 gallons would take you 126 miles. At 20 mpg, your distance traveled would be twice as far, or 252 miles. The reason we do approximate arithmetic is to obtain a quick check on the result without having to use a calculator.

fuel economy must be a little better than 20 mpg, and 21.7 mpg is a little better than 20 mpg. Thus, our answer does seem reasonable.

There are other ways we might have approximated the arithmetic to obtain a rough check on our answer. The point here is that you should look for ways to do simple arithmetic to obtain an approximate answer to the calculation so that you have a feel for whether the answer is of about the right size. With practice, this check will only take a short time and it will often save you from making dumb mistakes.

**Proportional Relationships**

In solving problems, we usually search for relationships between quantities. By *relationship*, we mean a connection of some sort between the quantities. For example, there is a relationship between how far we drive and the amount of fuel consumed. Many such relationships can be classified as either direct or inverse porportionalities.

In a direct proportionality, as one quantity increases by a certain factor (for example, it doubles), the other quantity increases by the same factor (it also doubles). Fuel usage falls into this category. If you double the distance you drive, you double the amount of fuel used. The relationship between distance traveled and fuel consumed is a direct proportionality[4], and we say that "the distance traveled is *directly proportional* to the amount of fuel consumed."

In an inverse proportionality, as one quantity *increases* by a certain factor, the other quantity *decreases* by the same factor. An example is the relationship between the speed you travel and the time required to go a given distance. If you go twice as fast, it takes you only half as long to get to your destination. Here we would say that "the time required to go a given distance is *inversely proportional* to speed at which you travel."

The two kinds of proportionalities we've just described are often useful in doing the mental arithmetic involved in checking whether an answer is correct. In fact, we used the direct proportionality between fuel consumed and distance traveled to check our answer to the fuel economy problem described above. We first rounded our answer to 20 mpg, and then noted that 1 gallon of fuel would take us 20 miles. Because of the direct proportionality, we realized that if we multiply the amount of fuel used by 12.6 (giving 12.6 gallons), we would multiply the distance traveled by 12.6 to give 252 miles.

**Mathematical Operations Involving Units**

As you know, when you multiply two fractions and find the same number in both the numerator and denominator, you can "cancel" the number and simplify the arithmetic.

$$\frac{\cancel{3}}{5} \times \frac{20}{\cancel{3}} = \frac{20}{5} = 4$$

The number 3 can be canceled from numerator and denominator

---

[4]As you probably know, fuel economy varies with the speed you drive, so to use this proportionality reliably, you would have to assume the same speed.

Also, when the same number is multiplied together two or more times, you can express the arithmetic as the number raised to a power (e.g., squared, cubed, etc.).

$$3 \times 3 = 3^2$$

$$3 \times 3 \times 3 = 3^3$$

A very useful property of units is that they undergo the same kinds of mathematical operations that numbers do. Thus, when the same unit appears in numerator and denominator, we can cancel it.

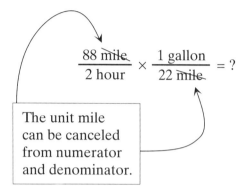

$$\frac{88 \cancel{\text{mile}}}{2 \text{ hour}} \times \frac{1 \text{ gallon}}{22 \cancel{\text{mile}}} = ?$$

The unit mile can be canceled from numerator and denominator.

After the units are canceled and the arithmetic is performed, the answer is $\frac{2 \text{ gallon}}{\text{hour}}$.

When the same unit is multiplied two or more times, we can write the unit raised to a power. For example,

$$2 \text{ ft} \times 3 \text{ ft} = 6 \text{ ft}^2$$

$$2 \text{ ft} \times 3 \text{ ft} \times 4 \text{ ft} = 24 \text{ ft}^3$$

**The Factor-Label Method and Conversion Factors**

The factor-label method is a procedure we use to set up the arithmetic correctly when solving a problem. It is based on the ability of units to cancel from numerator and denominator of a fraction, as described above. To apply this method, we view a problem as a conversion from some initial set of units to a final desired set of units. For example, if we need to know how many inches are in 1.5 ft, we begin by restating the problem as

$$1.5 \text{ ft} = ? \text{ in.}$$

To change the units from ft to in., we multiply the given quantity, 1.5 ft, by a conversion factor that will change the units from ft to in.

$$1.5 \text{ ft} \times \left( \frac{\text{conversion}}{\text{factor}} \right) = \text{answer in inches}$$

A **conversion factor** is a fraction that we form from a relationship between units. In this case, the relationship is between feet and inches, which can be expressed by the equation

$$1 \text{ ft} = 12 \text{ in.}$$

We can make a conversion factor by dividing both sides of the equation by 1 ft

$$\frac{1 \text{ ft}}{1 \text{ ft}} = \frac{12 \text{ in.}}{1 \text{ ft}}$$

Notice that the quantities in the numerator and denominator on the left cancel, so we can write

$$1 = \frac{\cancel{1 \text{ ft}}}{\cancel{1 \text{ ft}}} = \frac{12 \text{ in.}}{1 \text{ ft}}$$

Because the fraction 12 in./1 ft is numerically equivalent to 1, we can multiply something by it and not change its magnitude. (Multiplying something by 1 doesn't change its size.) Using this as our conversion factor, then, gives

$$1.5 \cancel{\text{ ft}} \times \frac{12 \text{ in.}}{1 \cancel{\text{ ft}}} = \text{answer in inches}$$

Notice that the units ft cancel[5], leaving us with units of inches, which are the units we want for the answer. Performing the arithmetic gives 18 in., which is the correct answer.

$$1.5 \cancel{\text{ ft}} \times \frac{12 \text{ in.}}{1 \cancel{\text{ ft}}} = 18 \text{ in.}$$

The relationship between feet and inches that we employed above can actually be used to form *two* different conversion factors. One is formed by dividing both sides by 1 ft, as we've done. The other is formed by dividing both sides by 12 in.

$$\frac{1 \text{ ft}}{12 \text{ in.}} = \frac{\cancel{12 \text{ in.}}}{\cancel{12 \text{ in.}}} = 1$$

Notice that if we were to use this second conversion factor in performing the conversion, the units "ft" do not cancel. Instead, the units work out to be ft$^2$/in., which make no sense. The answer, of course, is also incorrect.

$$1.5 \text{ ft} \times \frac{1 \text{ ft}}{12 \text{ in.}} = 0.12 \text{ ft}^2/\text{in.}$$

This illustrates one of the greatest benefits in using the factor-label method—if the problem is set up incorrectly, the units will not cancel properly to give the desired units of the answer. If the units are wrong, we can also be sure that the answer is wrong. Thus, the units have informed us that we've made an error and that we must rework the setup of the problem.

---

[5]A quantity such as 1.5 ft can be viewed as a fraction in which the denominator is 1.

$$1.5\text{ft} = \frac{1.5 \text{ ft}}{1}$$

This places the unit ft in the numerator where it can be canceled by the unit ft in the denominator of the conversion factor.

**Forming and Applying Conversion Factors**

In general, when we can express a relationship between units in the form of an equation, two conversion factors can be derived from it. For example, if we wished to convert between minutes and seconds, we would use the following.

$$60 \text{ s} = 1 \text{ min}$$

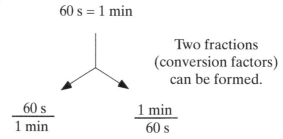

Two fractions (conversion factors) can be formed.

$$\frac{60 \text{ s}}{1 \text{ min}} \qquad \frac{1 \text{ min}}{60 \text{ s}}$$

The connection between feet and inches used earlier, and between minutes and seconds are both derived from definitions. Often we will use relationships that come to use in this way. We can also form conversion factors from relationships established by measurements. For instance, in an earlier example we described a situation in which we've determined that a car is able to travel 274 mile using 12.6 gallons of fuel. In making these measurements, we've established a relationship between distance traveled and fuel consumed that we can express in an equation format.

$$274 \text{ mile} \Leftrightarrow 12.6 \text{ gallon}$$

Notice, however, that we've used the symbol $\Leftrightarrow$ rather than an equals sign. This is because miles can't actually "equal" gallons; they're two different kinds of units. The symbol $\Leftrightarrow$ is read as "is equivalent to." In other words, for this car, "traveling 274 miles *is equivalent to* using 12.6 gallons of fuel." Although the symbol is different, it behaves like an equal sign when we wish to form conversion factors. Thus,

$$274 \text{ mile} \Leftrightarrow 12.6 \text{ gallon}$$

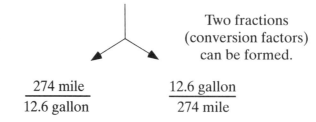

Two fractions (conversion factors) can be formed.

$$\frac{274 \text{ mile}}{12.6 \text{ gallon}} \qquad \frac{12.6 \text{ gallon}}{274 \text{ mile}}$$

Once you've learned how to form conversion factors, the next step is learning how to select the correct one for use in a calculation. Here we let the units be our guide. For instance, suppose we wished to know how many gallons of fuel we would need to travel 850 miles in the car described above. Let's begin by stating the problem in the form of an equation (or an equivalence, to be more exact)[6].

$$850 \text{ mile} \Leftrightarrow ? \text{ gallon}$$

---

[6]Don't be overly concerned about whether you use an equals sign or the symbol $\Leftrightarrow$. The setup of the problem and finding the answer follows the same route.

To find the answer by the factor-label method, we will multiply the 850 mile by a conversion factor to convert to gallons.

$$850 \text{ mile} \times \left( \begin{array}{c} \text{conversion} \\ \text{factor} \end{array} \right) \Leftrightarrow ? \text{ gallon}$$

The conversion factor we select has to eliminate the unit mile by cancellation. Only one of the two conversion factors does this, so that's the one we select.

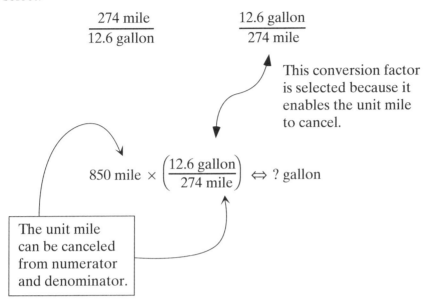

$$\frac{274 \text{ mile}}{12.6 \text{ gallon}} \qquad\qquad \frac{12.6 \text{ gallon}}{274 \text{ mile}}$$

This conversion factor is selected because it enables the unit mile to cancel.

$$850 \text{ mile} \times \left( \frac{12.6 \text{ gallon}}{274 \text{ mile}} \right) \Leftrightarrow ? \text{ gallon}$$

The unit mile can be canceled from numerator and denominator.

Once we've chosen the conversion factor, we cancel the units and perform the arithmetic.

$$850 \text{ \cancel{mile}} \times \left( \frac{12.6 \text{ gallon}}{274 \text{ \cancel{mile}}} \right) \Leftrightarrow 39.1 \text{ gallon}$$

Notice that if we had selected the other conversion factor, the units would not work out correctly.

The unit mile does not cancel.

$$850 \text{ mile} \times \left( \frac{274 \text{ mile}}{12.6 \text{ gallon}} \right) \Leftrightarrow ? \ \frac{\text{mile}^2}{\text{gallon}}$$

As you see in this example, the cancellation of units serves as a guide in selecting conversion factors as we set up the solution to the problem. In fact, the need to cancel particular units in arriving at the answer to a problem often serves as a clue to the kinds of information and relationships we have to find. For example, when we have arrived at the stage

$$850 \text{ mile} \times \left( \begin{array}{c} \text{conversion} \\ \text{factor} \end{array} \right) \Leftrightarrow ? \text{ gallon}$$

we would recognize that to obtain an answer we must have a relationship between gallons of fuel used and miles traveled. If this information were not already available, we would have to obtain it by some means before we could solve the problem.

## The Importance of Correct Relationships Between Units

In setting up problems using the factor-label method, it's essential that we use proper and correct relationships between the units. In converting between feet and inches, we need to use the relationship

$$1 \text{ ft} = 12 \text{ in.}$$

If we didn't know this relationship, we couldn't just make one up, such as

$$5 \text{ ft} = 24 \text{ in.}$$

Although we could use it to get the units to cancel, the answer is bound to be wrong because the relationship we've used in incorrect. Establishing correct relationships between units is one of the difficulties students often have in applying this approach. Before you form and apply a conversion factor, ask yourself "is this relationship between the units correct?"

## The Factor-Label Method and the "Toolbox" Approach

In our earlier discussion of how the toolbox concept can be applied to problem solving, we emphasized the importance of analyzing a problem so that critical links and the tools that are needed to obtain a solution can be identified. If the problem involves calculations, the tools are usually relationships between units, which can be stated in equation form. Once these statements have been assembled, the factor-label method provides guidance in assembling the arithmetic. By assuring the units cancel correctly, the factor-label method gives us confidence that we are doing the correct math. In a sense, then, the factor-label method is itself a tool that we use to assemble the arithmetic correctly. With this in mind now, let's take another look at a problem we worked on earlier to see how we might take a more formal problem solving approach. For convenience, we'll restate the problem.

### PROBLEM 1

During a cross country trip on Interstate I-80, a driver became concerned that her speedometer was inaccurate. To avoid a speeding ticket she decided to check the speedometer. She adjusted the speed of her car until the speedometer read 65 mph and set the car on speed control so the speed would remain constant. She then used the stopwatch feature of her wristwatch to time how long it took to travel exactly 5 miles, using the mile markers along the side of the road to measure her progress. She found that it took 4 minutes and 24 seconds to travel the five miles. She then pulled into a rest stop to calculate her actual speed over the five-mile stretch. What was her actual speed in miles per hour (mph)?

*Analysis:* The first, and most important step in solving a problem is the **analysis**. Here we study the data given, search for any critical links that might exist, and decide how we will proceed with the solution. We look for unit relationships (our tools), which we will use to solve the problem. After doing all this, we come to the easy part—setting up the **solution** and doing necessary calculations. We encourage you to think of problem solving as involving these two distinct stages: analysis first, followed by solution.

This problem involves a lot of words, so let's begin by extracting the critical information.

It took ***4 minutes and 24 seconds***

to travel ***5.00 miles***

How fast was she going, in ***miles per hour?***

As we've noted, the word "per" can be interpreted to mean "divided by," so our answer must have the units: mile "divided by" hour

$$\text{units of answer} = \frac{\text{mile}}{\text{hour}}$$

Understanding that we need to assemble the data in a way to give us this desired ratio of units is really our critical link in solving this problem. We see that we need to have distance in the numerator and time in the denominator. Let's use the data to set this up.

$$\text{speed} = \frac{5.00 \text{ mile}}{4 \text{ minutes and } 24 \text{ seconds}}$$

Now we can see that we're not too far from having the answer. We need to change the denominator into units of hour. Let's change 24 seconds to minutes, add the result to the 4 minutes, and then change the total minutes to hours. To do this, we need the relationships between seconds and minutes and between minutes and hours. (These are our tools.) We'll state them in equation form:

$$1 \text{ minute} = 60 \text{ second}$$
$$1 \text{ hour} = 60 \text{ minute}$$

Now that we know how we're going to proceed, we can work out the solution.

***Solution:*** First, we convert 24 seconds to minutes. Although this is really a simple exercise, let's illustrate how we perform the conversion using the factor-label method. We use the relationship 1 minute = 60 second to form a conversion factor that will cancel second and give the desired unit, minute.

$$24 \ \cancel{\text{second}} \times \frac{1 \text{ minute}}{60 \ \cancel{\text{second}}} = 0.40 \text{ minute}$$

The total time required to go five miles was therefore 4.40 minute. Now we convert this to hour using 1 hour = 60 minute. Once again, we form a conversion factor so units cancel correctly.

$$4.40 \ \cancel{\text{minute}} \times \frac{1 \text{ hour}}{60 \ \cancel{\text{minute}}} = 0.0733 \text{ hour}$$

Now we're able to calculate her speed by dividing the distance, 5.00 mile, by the time, 0.0733 hour.

$$\text{speed} = \frac{5.00 \text{ mile}}{0.0733 \text{ hour}} = 68.2 \frac{\text{mile}}{\text{hour}} = 68.2 \text{ mph}$$

The answer, therefore, is that the driver's speed was 68.2 mph.

***Is the Answer Reasonable?*** In any problem, this should always be your last step. We went through this already with this problem (page 6), and you might find it helpful to review our discussion.

---

**Stringing Conversion Factors Together**

The factor-label method is especially useful in multi-step calculations where two or more conversion factors are needed to go from the starting units to those of the answer. For example, suppose you wanted to know how may minutes there are in 2.00 weeks.

$$2.00 \text{ week} = ? \text{ minutes}$$

We could do this in a one step calculation if we knew how many minutes are in one week, but most people don't keep such numbers in their heads. However, we do know how many days there are in a week, and we know how many hours there are in a day and how many minutes in an hour.

$$1 \text{ week} = 7 \text{ day}$$
$$1 \text{ day} = 24 \text{ hour}$$
$$1 \text{ hour} = 60 \text{ minute}$$

We can apply these relationships one after another, being sure the units cancel, until we have the final desired unit, minute

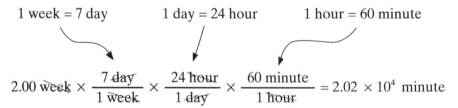

When we do this, we say that we are "stringing together" the conversion factors.

---

**Following the Path Through the Units**

In a multi-step calculation, a way to be sure you have all the information necessary to set up the string of conversion factors is to follow the "path" through the units, starting with the units given and going through until you reach the units of the answer. For instance, in the list of relationships we used in the preceding example,

$$1 \text{ week} = 7 \text{ day}$$
$$1 \text{ day} = 24 \text{ hour}$$
$$1 \text{ hour} = 60 \text{ minute}$$

we can follow the path: week → day → hour → minute. Thus, the first equation takes us from *week* to *day*, the second from *day* to *hour*, and the third from *hour* to *minute*. If one of these relationships were missing (for example, the middle one), the path would be interrupted.

We can connect week to day and hour to minute, but we are missing the connection between day and hour

1 week = 7 day

1 hour = 60 minute

week → day $\overset{?}{-}$ hour → minute

We see that to solve the problem, we need an additional relationship that connects the units day and hour.

---

**Additional Problems**

1. The 88-mile long West Virginia Turnpike cost about $683 million dollars to build. How much did the turnpike cost per inch?

2. A "significant emission rate" for sulfur dioxide emissions under the Clean Air Act is 40.0 tons $SO_2$ per year. A coal-fired power plant in Western Maryland burns bituminous coal containing 2.5% sulfur (2.5 tons of sulfur per hundred tons of coal). Burning 1.0 ton of sulfur produces 2.0 tons of $SO_2$. What is the maximum number of tons of coal the plant can burn each year before it is considered a significant source of $SO_2$ under the Clean Air Act?

3. Americium (chemical symbol Am) is an artificially produced element used in smoke detectors. A dose of as little as 0.02 μg of americium can produce toxic effects. How many ounces of Am is this? (1 lb = 453.6 g; 1 lb = 16 oz; 1 μg = $10^{-6}$ g)

4. In 1930, Albert Einstein sold his autographs for $3.00 apiece to raise funds for the poor in Berlin. If Einstein could write an autograph every 2.00 seconds, how many hours would it take to raise one million dollars?

5. Saffron is an expensive spice ($13.06 for a packet containing 0.060 oz of saffron), made from the stamens of the autumn crocus (Crocus sativa). Each flower on the plant yields 4 stamens. It takes about 250,000 stamens to make a pound of saffron. Each plant needs about one square foot of land and produces about 4 harvestable flowers per season. How many square feet must be planted with autumn crocuses to produce a million dollars worth of saffron?

1. $122/inch, 2. 800 tons coal, 3. $7 \times 10^{-10}$ oz, 4. 185 hours, 5. $4.5 \times 10^6$ **Answers**
square feet

# 2. SI UNITS AND CONVERSIONS

Now that we've discussed with you some of the essential elements of problem solving, we turn our attention to problems related to chemistry. We begin with unit conversion and will concentrate on the units we use in the sciences, which are called SI units. These are described in the textbook in Section 1.4, which begins on page 9. For now, the most important of the base units (Table 1.1, page 10) are those for length (meter, m), mass (kilogram, kg), time (second, s), temperature (kelvin, K), and amount of substance (mole, mol). Be sure you know both the names and the symbols for these units. (The unit for electric current (ampere, A) will be important in discussions in Chapter 19, but the unit for luminous intensity isn't used in the text.)

## SI Prefixes–Relating SI Units

**Skills Analysis**

1. About $9 \times 10^9$ liters of bottled water are consumed worldwide each year. How many gigaliters of water of is this?

2. How many micrometers are in one meter?

3. Suppose 5100 milliseconds pass before thunder is heard after a flash of lightning. If sound travels at a rate of 331 m/s under prevailing conditions, how far away was the lightning flash, in km?

4. The largest viruses have a diameter of about 250 nm. How many inches is this?

**Answers**

**1.** 9 GL. If you missed this question, you need to study this entire section and memorize the list of SI prefixes printed in red in Table 1.2 on page 11 of the text. **2.** 1 m $= 10^6$ μm. If you missed this question, study the entire section, paying special attention to the example given under "Writing Relationships between Base Units and Modified Units." **3.** 1.7 km. If you missed this question, study the entire section, paying special attention to the examples given under "Conversions Among SI Units." **4.** $9.8 \times 10^{-6}$ in. If you missed this question, study the entire section, especially the examples given under "SI to English Conversions."

**Modifying SI Base Units**

In the textbook it was noted that the base units of the SI are often of an inconvenient size. For example, the thickness of a human hair is about 0.0002 m and an atom of carbon has a diameter of about 0.0000000015 m. These small numbers are difficult to work with and to comprehend because of all the zeros. To form units of a more convenient size, the SI applies modifiers to the base units to give either larger or smaller units. These modifiers are powers of ten, and the name of the modified unit is formed by adding a prefix, called an SI prefix, to the name of the base unit. An example is the prefix centi-, which indicates a modifier of $10^{-2}$. Thus, a *centi*meter is $10^{-2}$ m, or 0.01 m. The symbol for the modified unit is formed by placing a symbol prefix in front of the symbol for the base unit. The symbol for centi- is c and the symbol for meter is m, so a centimeter is cm.

> The symbol for centi- is c and the symbol for meter is m. Put them together and you have the symbol for centimeter (cm).

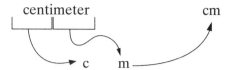

Table 1.2 on page 11 of the textbook contains a list of the SI prefixes, their symbols, and the factors by which they modify a base unit. The most important ones are printed in red and it is essential that you learn (memorize) them. You should also have a qualitative feel for the relative sizes of the modified units relative to the base units. Those SI prefixes that correspond to multiplying factors of 10 to a negative power give modified units that are smaller than the base unit. If the multiplying factor is 10 to a positive power, then the modified unit is larger than the base unit.

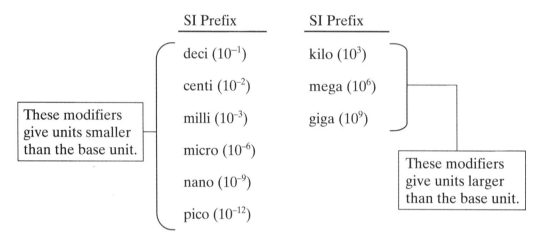

| SI Prefix | SI Prefix |
| --- | --- |
| deci ($10^{-1}$) | kilo ($10^{3}$) |
| centi ($10^{-2}$) | mega ($10^{6}$) |
| milli ($10^{-3}$) | giga ($10^{9}$) |
| micro ($10^{-6}$) | |
| nano ($10^{-9}$) | |
| pico ($10^{-12}$) | |

These modifiers give units smaller than the base unit.

These modifiers give units larger than the base unit.

Thus, a decimeter and a millimeter are both smaller units than a meter, and a kilometer is larger than a meter.

**Writing Relationships Between Base Units and Modified Units**

There will be many times when you will find it necessary to convert between base units and modified units, for example, between meters and centimeters. To do this you must be able to write the correct relationship between them, a process that is very simple as long as you remember the definitions for the SI prefixes. For example, to write the relationship between centimeter (cm) and meter (m), recall that centi- *means* $10^{-2}$. Therefore, just substitute $10^{-2}$ for the prefix centi (c).

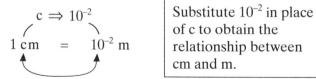

> Substitute $10^{-2}$ in place of c to obtain the relationship between cm and m.

Let's look an a sample problem that requires this kind of operation.

**PROBLEM 2**
How many nanometers are there in 1 meter?

*Analysis:* The problem is actually a little deceptive, and you might be tempted to jump at the answer. The critical link in solving this problem is correctly stating the relationship between nanometer (nm) and meter (m). We do this by writing the definition of a nanometer in terms of the base unit, meter. In other words, we find the value for the question mark in the expression

$$1 \text{ nm} = ? \text{ m}$$

Once we've done this, we can use the relationship to form a conversion factor that will allow us to find the number of nanometers in one meter.

$$1 \text{ m} \times \left( \frac{\text{conversion}}{\text{factor}} \right) = ? \text{ nm}$$

*Solution:* From Table 1.2, we recall that nano- means $10^{-9}$, so following the example we worked for the relationship between centimeter and meter, we substitute $10^{-9}$ for "n."

$$1 \text{ nm} \quad = \quad ? \text{ m}$$

with $10^{-9}$ shown over the equals relationship

$$\Downarrow$$

$$1 \text{ nm} \quad = \quad 10^{-9} \text{ m}$$

This is the correct relationship between nanometer and meter, which we can use as a tool whenever we want to convert between these two units. To find the number of nanometers in one meter, we form the conversion factor so that the unit meter cancels.

$$1 \text{ m} \times \frac{1 \text{ nm}}{10^{-9} \text{ m}} = 10^9 \text{ nm}$$

Thus, 1 m consists of $10^9$ nm.

*Is the Answer Reasonable?* Earlier we said that the problem is somewhat deceptive in the way it is stated. If you focus only on nano meaning $10^{-9}$ and are a bit careless, you might have been tempted to write: $1 \text{ m} = 10^{-9}$ nm. You can recognize your error, however, if you keep in mind what the SI prefix does. The prefix nano- gives a unit that's *smaller* than the base unit, so a nanometer is smaller than a meter (in fact, a whole lot smaller). Because the nanometer is so small, there must be a large number of them in one meter. The answer, $10^9$, is a large number, so the answer is reasonable. However, an answer of $1 \text{ m} = 10^{-9}$ nm doesn't make sense because it suggests that a meter is a small fraction of a nanometer.

**Conversions Among SI Units**

Let's look now at some unit conversion problems in which we will use the SI prefixes as tools to perform the calculations.

### PROBLEM 3

If a bullet fired from a gun travels at a speed of $1.6 \times 10^3$ kilometers per hour (km/hr), what distance (expressed in centimeters) does the bullet travel in 2.0 milliseconds?

*Analysis:* Let's begin by studying what we are being asked to find. The problem asks us to find the *distance* (in cm) that the bullet travels in a certain *time* (2.0 milliseconds, or 2.0 ms[7]). Let's express this in the form of an equivalence.

$$? \text{ cm} \Leftrightarrow 2.0 \text{ ms}$$

or, placing the given quantity on the left,

$$2.0 \text{ ms} \Leftrightarrow ? \text{ cm}$$

Unit conversion problems can almost always be solved using the factor-label method, so we anticipate that we will multiply 2.0 ms by one or more conversion factors to obtain our answer.

$$2.0 \text{ ms} \times \left( \frac{\text{one or more}}{\text{conversion factors}} \right) \Leftrightarrow \text{answer in cm}$$

The tools we need to form the conversion factors are the relationships between units. One of these must relate distance and time, and in the problem we are given the speed in km/hr. This tells us that in 1 hr, the bullet would travel $1.6 \times 10^3$ km. Let's write that down as follows:

$$1 \text{ hr} \Leftrightarrow 1.6 \times 10^3 \text{ km}$$

To complete a path through the units, we need a series of unit relationships that will take us from milliseconds to hour and then another series of that will take us from kilometer to centimeter.

Let's first go from hour to milliseconds. In 1 hour there are 3600 seconds.

$$1 \text{ hr} = 3600 \text{ s}$$

and 1 millisecond is $10^{-3}$ second.

$$1 \text{ ms} = 10^{-3} \text{ s}$$

Now let's look at the relationships that take us from kilometers to centimeters. The SI prefixes tell us the following:

$$1 \text{ km} = 10^3 \text{ m}$$
$$1 \text{ cm} = 10^{-2} \text{ m}$$

---

[7]The symbol for milli is m; the symbol for second is s. Therefore, the symbol for millisecond is ms.

20   Analyzing and Solving Chemistry Problems

To better see the flow of units, let's rearrange these five unit relation-ships as follows, with the unit of the given quantity on top and the unit of the desired answer at the bottom:

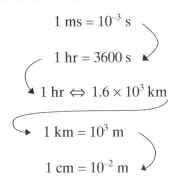

$$1 \text{ ms} = 10^{-3} \text{ s}$$

$$1 \text{ hr} = 3600 \text{ s}$$

$$1 \text{ hr} \Leftrightarrow 1.6 \times 10^3 \text{ km}$$

$$1 \text{ km} = 10^3 \text{ m}$$

$$1 \text{ cm} = 10^{-2} \text{ m}$$

Notice that we now have an unbroken path that will take us from the unit ms to the unit cm. We're now ready to complete the solution.

***Solution:*** This is really the easy part of the problem. All we need to do is form conversion factors, being sure that the units cancel properly to give those of the answer, and then do the arithmetic. (Add the can-cel marks yourself, just to be sure the units cancel.)

$$2.0 \text{ ms} \times \frac{10^{-3} \text{ s}}{1 \text{ ms}} \times \frac{1 \text{ hr}}{3600 \text{ s}} \times \frac{1.6 \times 10^3 \text{ km}}{1 \text{ hr}} \times \frac{10^3 \text{ m}}{1 \text{ km}} \times \frac{1 \text{ cm}}{10^{-2} \text{ m}} = 89 \text{ cm}$$

The result of our calculation tells us that in 2.0 milliseconds, the bullet will travel a distance of 89 cm. (Notice that we've rounded the answer to two significant figures.)

***Is the Answer Reasonable?*** Because of all of the conversion factors involved and the mix of large and small numbers, doing approximate arithmetic would be tedious. However, we do know that a bullet trav-els pretty fast, and it does seem reasonable that in 2.0 milliseconds (two thousandths of a second) the bullet would travel almost a meter, or about 3 feet.

The following problem is similar to the preceding one, but requires just a bit more thinking.

## PROBLEM 4
Suppose a photographer takes a super high-speed flash photo in a sta-dium that is 300 m across and a person 200 m away observes the flash. If the duration of the flash is 1.00 microsecond, will the flash be over before the person 200 m away begins to observe it? The speed of light is $3.00 \times 10^8 \text{m s}^{-1}$.

***Analysis:*** What makes this problem interesting is that we have to do some thinking to decide what it is we are going to have to calcu-late. We know the light from the flash will travel 200 m from the camera to the observer and we are told how fast light travels. The question is, will the light from the flash reach the observer in less than a microsecond. If so, then the light given off at the beginning of the flash will be observed before the flash has finished. If it takes longer than a microsecond for the light to travel 200 m, then the flash will be over before any of the light reaches the observer. What we need to calculate therefore is how long it takes (in microseconds, μs) for the light to travel 200 m. Let's write this in the form of an equivalence.

$$200 \text{ m} \Leftrightarrow \text{? } \mu\text{s}$$

The speed of light has units m s$^{-1}$, which means m/s. It tells us that in 1 s light travels $3.00 \times 10^8$ m.

$$1 \text{ s} \Leftrightarrow 3.00 \times 10^8 \text{ m}$$

To finish the problem we need the relationship between second and microsecond, which we obtain from the definition of the SI prefix micro-.

$$1 \text{ μs} = 10^{-6} \text{ s}$$

We now have a path through the units from the given unit, m, to the desired unit, μs.

$$1 \text{ s} \Leftrightarrow 3.00 \times 10^8 \text{ m}$$
$$1 \text{ μs} = 10^{-6} \text{ s}$$

***Solution:*** To find the time required for light to travel 200 m, we form conversion factors as follows (once again, add cancel marks yourself to see how the units cancel):

$$200 \text{ m} \times \frac{1 \text{ s}}{3.00 \times 10^8 \text{ m}} \times \frac{1 \text{ μs}}{10^{-6} \text{ s}} = 0.667 \text{ μs}$$

It takes less than a microsecond (millionth of a second) for the light to reach the observer, so when the first light given off by the flash reaches the observer, the flash will still be in progress.

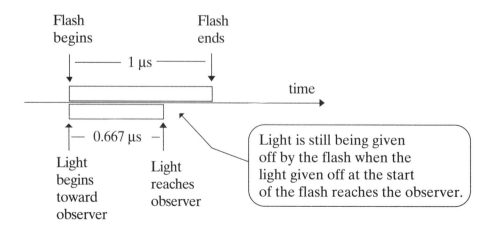

***Is the Answer Reasonable?*** If we check the arithmetic, we find the answer is computed correctly, but does the answer *make sense*? We do know that light travels extremely fast, so it seems reasonable that it could take less than a millionth of a second for the light to travel to the observer. The answer does seem reasonable because it agrees with what we know about how fast light travels.

In the United States we commonly use English units in our daily lives, but in the laboratory SI units are the norm. It is only rarely that we have to make conversions between these two sets of units, but occasionally such calculations are needed. There are many conversion relationships available in tables, some of which are found in Table 1.3 on page 12 of the textbook, but it is not really necessary to memorize all of them. As long as you can make conversion within each system of units, you need only a few bridges between the two systems. The ones you elect to learn is up to you, but we recommend the following:

$$\text{Length:} \quad 1 \text{ in.} = 2.54 \text{ cm}$$
$$\text{Mass:} \quad 1 \text{ lb} = 454 \text{ g}$$
$$\text{Volume:} \quad 1 \text{ qt} = 946 \text{ mL}$$

Notice that we've expressed them to three significant figures. (If your calculations require more than this, then it's time to refer to a table of English–Metric conversions.) Problems that require these conversions are very similar to the unit conversion problems we've already examined.

## PROBLEM 5

A human hair has a thickness of about 200 $\mu$m (assume one significant figure). What is this expressed in units of inches?

***Analysis:*** As before, let's begin by expressing the question in terms of an equation:

$$200 \ \mu\text{m} = ? \text{ in.}$$

The tools we need to solve the problem are the unit relationships. We're working with a Metric–English conversion, so the critical link in solving the problem is selecting the conversion between the two systems. We're working with length units, so the relationship we will use is

$$1 \text{ in.} = 2.54 \text{ cm}$$

Now we can see that to complete the path through the units, we need relationships that will take us from $\mu$m (the units of the given quantity) to cm. As before, we obtain them from the definitions of the SI prefixes.

$$1 \ \mu\text{m} = 10^{-6} \text{ m}$$
$$1 \text{ cm} = 10^{-2} \text{ m}$$

Just to see how the units flow, let's rearrange the conversions[8].

$$1 \, \mu m = 10^{-6} \, m$$
$$1 \, cm = 10^{-2} \, m$$
$$1 \, in. = 2.54 \, cm$$

Notice once again that we can follow a path through the units from those at the start ($\mu m$) to those at the end (in.).

**Solution:** The hard work is done. Now we just apply the conversions as conversion factors and calculate the answer. (Add the cancel marks yourself.)

$$200 \, \mu m \times \frac{10^{-6} \, m}{1 \, \mu m} \times \frac{1 \, cm}{10^{-2} \, m} \times \frac{1 \, in.}{2.54 \, cm} = 8 \times 10^{-3} \, in.$$

**Is the Answer Reasonable?** In decimal form, the answer is 0.008 in. (that's eight thousandths of an inch). A hair is certainly pretty thin, so the answer does seem reasonable.

**Additional Problems**

1. The Near Earth Asteroid Rendezvous spacecraft was recently spotted through an optical telescope, making it the most distant manmade object ever seen. The craft was 33.6 million kilometers away. How many gigameters is this?

2. The silicon needles on new blood analyzing devices can painlessly collect samples of blood with volumes of about 100 nanoliters. How many samples of this size would fit in a conventional 5 mL hypodermic syringe?

3. The largest cell in the human body is the ovum, which is about 1/175th of an inch in diameter. How many millimeters is this?

4. The smallest viruses are about 20 nm across. If the viruses are placed in a straight line, side by side and just touching, how many would stretch across the space of an inch?

**Answers**

1. 33.6 Gm, 2. 50,000 samples, 3. 0.145 mm, 4. $1.3 \times 10^6$ viruses

**Relating Length, Area, and Volume**

**Skills Analysis**

1. Write a conversion factor that converts cubic inches to cubic centimeters.

2. What is the volume of an aquarium (in $cm^3$) that is 1.0 m wide, 3.0 m long, and 2.0 m deep?

3. The earth has a surface area of about 510 square megameters ($5.1 \times 10^2 \, Mm^2$). What is this area in square meters ($m^2$)?

---

[8]In solving a problem, this step is not really necessary as long as you can follow the path through the units. We are rearranging the conversions just to make it easier for you to see the flow of units.

**4.** The lungs of a typical adult have an interior surface area of about $1.0 \times 10^2$ m$^2$. What is this area in square feet? (1 m = 39.37 in.; 12 in. = 1 ft)

If you miss any of these questions, you should study this entire section.

**1.** $\dfrac{16.38706 \text{ cm}^3}{1 \text{ in.}^3}$, **2.** $6.0 \times 10^6$ cm$^3$, **3.** $5.1 \times 10^{14}$ m$^2$, **4.** $1.1 \times 10^3$ ft$^2$

---

To determine the area of a rectangular object, we make two measurements—its length and its width. The product of these two gives the area. For example, in the figure below, the area of the object is the product of 5.0 cm and 3.0 cm.

**Calculating Area and Volume**

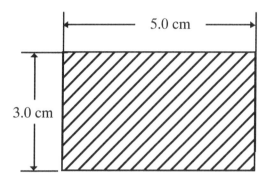

Area = 5.0 cm × 3.0 cm

= 15 cm$^2$

Notice that we multiply the numbers together and we also multiply the units together. The resulting unit, cm$^2$, can be given in words either as "square centimeter" or "centimeter squared." In general, the unit for area is a length unit squared (e.g., cm$^2$, m$^2$, ft$^2$, in.$^2$).

To measure volume, three dimensions are determined. For the rectangular box below, we measure its width, depth and height. The volume is a product of the three measurements.

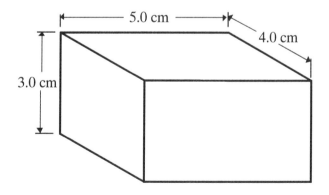

Volume = 5.0 cm × 4.0 cm × 3.0 cm

= 60 cm$^3$

Once again, the numbers are multiplied and so are the units. Expressed in words, the unit for area here is "cubic centimeter" or "centimeter cubed." In general, the unit for volume is a length-unit cubed (e.g., cm$^3$, m$^3$, ft$^3$, in.$^3$).

**Creating Relationships Between Squared or Cubed Units**

The SI does not define units for area or volume. Instead, we derive them by multiplying length units. Similarly, when we perform unit conversions, we have to derive the relationships between squared or cubed units. This is quite simple to do. For example, suppose we needed to convert between square centimeters ($cm^2$) and square meters ($m^2$). We begin with the relationship between meter and centimeter.

$$1 \text{ cm} = 10^{-2} \text{ m}$$

We then square both sides of the equation.

$$(1 \text{ cm})^2 = (10^{-2} \text{ m})^2$$

$$(1^2)(cm)^2 = (10^{-2})^2(m)^2$$

$$1 \text{ cm}^2 = 10^{-4} \text{ m}^2$$

It is very important to note that we *square both the number **and** the unit*!

When we need to find the relationship between cubed units, we begin with the basic relationship between the units and then cube both sides of the equation, being careful to cube *both* the numbers and the units. Thus, to find how cubic centimeters relate to cubic meters, we start with the SI definition of the centimeter.

$$1 \text{ cm} = 10^{-2} \text{ m}$$

Then we cube both sides of the equation.

$$(1 \text{ cm})^3 = (10^{-2} \text{ m})^3$$

$$(1^3)(cm)^3 = (10^{-2})^3(m)^3$$

$$1 \text{ cm}^3 = 10^{-6} \text{ m}^3$$

Remember, in forming these relationships you must square or cube both the number and the unit. Now let's look at an example that requires using volume units.

**PROBLEM 6**

What is the volume in cubic millimeters ($mm^3$) of the 60 $cm^3$ object described above?

*Analysis:* As we've done before, let's start by expressing the problem in the form of an equation.

$$60 \text{ cm}^3 = ? \text{ mm}^3$$

If we can find a path through the uncubed units from cm to mm, we can then form the necessary relationships to solve the problem by cubing each side of each conversion. Our basic tools, then, come from the definitions of cm and mm.

$$1 \text{ cm} = 10^{-2} \text{ m}$$

$$1 \text{ mm} = 10^{-3} \text{ m}$$

Note that these give us a path from cm to mm (cm → m → mm). Now we simply cube both sides of each equation to obtain the conversions we need. Again, we are careful to cube both the numbers and the units.

$$1 \text{ cm} = 10^{-2} \text{ m}$$

$$1 \text{ cm}^3 = 10^{-6} \text{ m}^3$$

and

$$1 \text{ mm} = 10^{-3} \text{ m}$$

$$1 \text{ mm}^3 = 10^{-9} \text{ m}^3$$

This gives us a path through the cubed units from cm³ to mm³.

$$1 \text{ cm}^3 = 10^{-6} \text{ m}^3$$

$$1 \text{ mm}^3 = 10^{-9} \text{ m}^3$$

**Solution:** Now we're ready to apply the conversions to solving the problem. As usual, we take care to be sure the units cancel, so we can be confident we're doing the correct arithmetic. (Add the cancel marks yourself.)

$$60 \text{ cm}^3 \times \frac{10^{-6} \text{ m}^3}{1 \text{ cm}^3} \times \frac{1 \text{ mm}^3}{10^{-9} \text{ m}^3} = 6.0 \times 10^4 \text{ mm}^3$$

**Is the Answer Reasonable?** This is a difficult question to answer quickly. A millimeter is smaller than a centimeter. In fact, one centimeter equals 10 mm (a fact that is useful to know, and which you can verify if you have a ruler with a centimeter/millimeter scale on it). One square centimeter is a square that is one centimeter (10 mm) on each side.

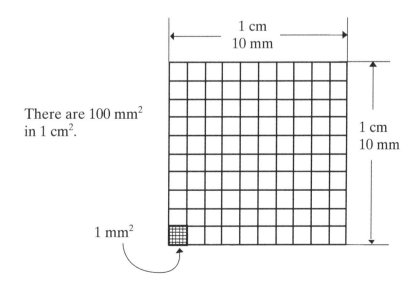

There are 100 mm² in 1 cm².

Each box is one square millimeter, and there are a hundred of them in one square centimeter. A cubic centimeter is a cube that is one centimeter (10 mm) along each edge, as shown below.

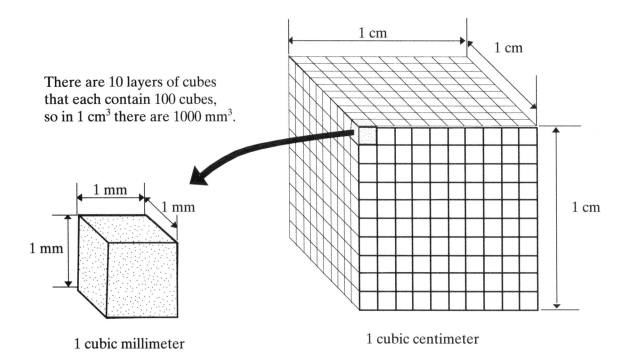

There are 10 layers of cubes that each contain 100 cubes, so in 1 cm³ there are 1000 mm³.

1 cubic millimeter

1 cubic centimeter

We can see that in 1 cm³ there are 1000 mm³. Therefore, in 60 cm³ there are 60,000 mm³, or $6.0 \times 10^4$ mm³ when written in scientific notation to express the correct number of significant figures.

This analysis is probably a lot more than you would want to do, but it shows how we can reason through the problem to find the answer. Clearly, the use of the factor-label method and the appropriate relationships among the cubed units is a lot simpler.

**Additional Problems**

1. Researchers at the University of New Orleans predict that the city of New Orleans will be completely underwater by the year 2100 if the current land loss rate ($6.5 \times 10^7$ square meters per year) continues in Louisiana. How many square miles of land will be lost over a 25 year period at this rate? (39.37 in. = 1 m; 12 in. = 1 ft; 5280 ft = 1 mi)

2. The body mass index (BMI) is calculated by dividing a person's mass by their height squared. The National Heart, Lung, and Blood Institute considers a person "overweight" if their BMI is higher than 25 kg/m², and "obese" if their BMI is 30 kg/m² or above. Would a person who is 5 feet 5 inches tall and weighs 157 pounds be considered overweight by these definitions? (2.202 lb = 1 kg; 1 m = 39.37 in.; 12 in. = 1 ft)

3. When fatty acids are dropped onto water, they spread across the surface to form a "monolayer"—a layer just one molecule thick. Reservoirs in dry countries are deliberately covered with fatty acid monolayers to slow evaporation rates. If each fatty acid molecule covers an area of 0.21 nm², how many molecules will cover the surface of a lake with a surface area of 10.0 km²?

## Relating Mass and Volume for a Single Substance

**1.** A cube of clay that measures 1.0 cm on a side weighs 5.00 g. What is the density of the clay?

**2.** Seawater has a density of 1.02 g/mL. Can this information be used to compute the volume of an iceberg that weighs 1,000,000 kg?

**3.** Mercury has a density of 13.6 g/mL. What is the mass of 1.00 L of mercury, in kg?

**1.** 5.0 g/cm$^3$. Read this entire section if you missed this one. **2.** No. A floating object displaces a mass of water equal to its own mass, so the iceberg displaces 1,000,000 kg of seawater. But a part of an iceberg is above water, so to calculate the volume of the iceberg, we need the density of the iceberg. Read the entire section if you missed this question, particularly "Using Density as a Conversion Factor." **3.** 13.6 kg Hg. Read the entire section if you missed this one. You may also want to review the section on SI prefixes.

**Density as a Critical Link**

We've now spent a lot of time discussing unit conversion, and if you have truly mastered these topics, you will see how well they apply to problems that deal more specifically with chemistry.

Among the physical properties of substances that are discussed in the textbook, density is one of the most useful (see pages 26 and 27). We calculate density ($d$) as a ratio of an object's mass ($m$) to its volume ($V$).

$$d = \frac{m}{V}$$

Typically, densities of liquids and solids have units of grams per cubic centimeter, g/cm$^3$ or g cm$^{-3}$. Gases, which are much less dense than liquids or solids, often have their densities reported in grams per liter, g/L or g L$^{-1}$.

Density is a useful property because it aids in the identification of substances; each substance has its own characteristic density. *Density also serves as the* **critical link** *in problems that ask us to relate the mass of a substance to its volume, or vice versa.* In other words, if you're working on a problem and need to relate the mass of a substance to its volume, or the volume to its mass, the tool you need to use is the density.

**Using Density as a Conversion Factor**

To use density as a tool in calculations, we have to know how to form conversion factors from it. As an example, let's consider the metal silver, which has a density of 10.5 g cm$^{-3}$ (Table 1.4, textbook page 27). The density tells us the mass of one cubic centimeter of the metal. This means we can write the following,

$$1 \text{ cm}^3 \text{ silver} = 10.5 \text{ g silver}$$

(Notice that we have included the word silver on both sides of this equation. When we use density, the $cm^3$ and g must both refer to the same substance.) We can use this relationship to form two conversion factors.

$$\frac{1 \text{ cm}^3 \text{ silver}}{10.5 \text{ g silver}} \quad \text{and} \quad \frac{10.5 \text{ g silver}}{1 \text{ cm}^3 \text{ silver}}$$

We would use the first factor if we want to cancel the unit "g silver." The second would be used if we want to cancel the unit "$cm^3$ silver."

**When Density Cannot Be Used to Relate Mass and Volume**

There are two criteria to using density to relate mass and volume. First, the density, mass, and volume must all refer to the same substance. Second, they should both refer to the same physical state. For example, we can't use the density of liquid water to find the volume of certain mass of gaseous water (steam).

### PROBLEM 7
What is the mass in grams of 12.4 $cm^3$ of silver?

*Analysis:* Let's begin by stating the problem in the form of an equation.

$$12.4 \text{ cm}^3 \text{ silver} = ? \text{ g silver}$$

The problem requires us to relate the mass and volume of the same substance, silver. The appropriate tool to accomplish this is the density, which we learned is 10.5 g $cm^{-3}$ for silver. To convert from $cm^3$ to g, we need to use the conversion factor on the right above.

*Solution:* Applying the conversion factor gives us the mass of silver.

$$12.4 \text{ cm}^3 \text{ silver} \times \frac{10.5 \text{ g silver}}{1 \text{ cm}^3 \text{ silver}} = 130 \text{ g silver}$$

*Is the Answer Reasonable?* The density tells us that each cubic centimeter of silver has a mass of about 10 g, so 12.4 $cm^3$ should have a mass of about 124 g. The answer is near this value, so it seems to be reasonable.

As we conclude this section on unit conversions, let's look at a problem that brings together a number the topics we've discussed. We will work through the problem carefully so you can follow all the reasoning steps involved.

### PROBLEM 8
An artist's statue has a surface area of about 15 $ft^2$. The artist plans to apply a gold plate coating to the statue and wants the coating to be 25 micrometers[9] thick. How many grams of gold will be required to give this gold coating to the statue?

*Analysis:* If we knew the density of gold (which we can look up in Table 1.4), we could calculate the mass of gold from the volume of

---

[9]A micrometer (μm) is also called a micron.

gold that will be applied to the statue. The volume of gold can be calculated by multiplying the area of the coating by its thickness.

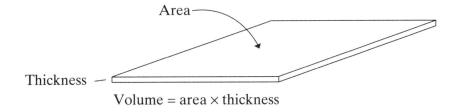

Volume = area × thickness

In Table 1.4 (textbook page 27) we find the density of gold given as $19.3 \text{ g cm}^{-3}$. To use it in a calculation, we need to find the volume of gold in cubic centimeters, We have the area in $\text{ft}^2$, which we will have to convert to $\text{cm}^2$. We will also have to convert the thickness from micrometers to centimeters. Then we can multiply the area (in $\text{cm}^2$) by the thickness (in cm) to obtain the volume in $\text{cm}^3$. Let's work on the area and assemble the tools we need to convert from $\text{ft}^2$ to $\text{cm}^2$.

First we will assemble the relationships to go from ft to cm. Then we will square each side to obtain the relationships among the squared units. We are going to need a conversion from English to metric length units, so we'll begin with

$$1 \text{ in.} = 2.54 \text{ cm}$$

To go from feet to inches we have

$$12 \text{ in.} = 1 \text{ ft}$$

Now we have a path from ft to cm.

$$12 \text{ in.} = 1 \text{ ft}$$
$$\downarrow$$
$$1 \text{ in.} = 2.54 \text{ cm}$$

Squaring each side gives the conversions we need to go from $\text{ft}^2$ to $\text{cm}^2$ (we've placed them in boxes).

$$(12 \text{ in.})^2 = (1 \text{ ft})^2$$

$$\boxed{144 \text{ in.}^2 = 1 \text{ ft}^2}$$

$$(1 \text{ in.})^2 = (2.54 \text{ cm})^2$$

$$\boxed{1 \text{ in.}^2 = 6.45 \text{ cm}^2}$$

We also need to convert between micrometer and centimeter. These are both SI units, so we can use the definitions of the SI prefixes.

$$1 \text{ }\mu\text{m} = 10^{-6} \text{ m}$$
$$1 \text{ cm} = 10^{-2} \text{ m}$$

Now we've got all the information we need to solve the problem.

***Solution:*** First, we'll calculate the area in cm².

$$15 \text{ ft}^2 = ? \text{ cm}^2$$

The relationships we'll use are those in boxes on page 31.

$$144 \text{ in.}^2 = 1 \text{ ft}^2$$

$$1 \text{ in.}^2 = 6.45 \text{ cm}^2$$

Assembling the conversion factors gives

$$15 \text{ ft}^2 \times \frac{144 \text{ in.}^2}{1 \text{ ft}^2} \times \frac{6.45 \text{ cm}^2}{1 \text{ in.}^2} = 1.4 \times 10^4 \text{ cm}^2 = \text{area}$$

To find the thickness in cm, we can start with the statement

$$25 \text{ } \mu\text{m} = ? \text{ cm}$$

Then we use the following relationships to form conversion factors.

$$1 \text{ } \mu\text{m} = 10^{-6} \text{ m}$$

$$1 \text{ cm} = 10^{-2} \text{ m}$$

The calculation becomes

$$25 \text{ } \mu\text{m} \times \frac{10^{-6} \text{ m}}{1 \text{ } \mu\text{m}} \times \frac{1 \text{ cm}}{10^{-2} \text{ m}} = 2.5 \times 10^{-3} \text{ cm} = \text{thickness}$$

The volume of the gold is the product of the area ($1.4 \times 10^4$ cm²) and the thickness ($2.5 \times 10^{-3}$ cm).

$$\text{Volume of gold} = (1.4 \times 10^4 \text{ cm}^2) \times (2.5 \times 10^{-3} \text{ cm})$$

$$= 34 \text{ cm}^3 \left(\begin{array}{c} \text{rounded to 2} \\ \text{significant figures} \end{array}\right)$$

Now that we have the volume of gold, we can use its density (19.3 g cm⁻³) to find the mass of gold. The density tells us the mass of 1 cm³ of gold.

$$1 \text{ cm}^3 \text{ gold} = 19.3 \text{ g gold}$$

We want to find the mass of 34 cm³ of gold.

$$34 \text{ cm}^3 \text{ gold} = ? \text{ g gold}$$

We use the conversion factor that allows us to cancel cm³ gold.

$$34 \text{ cm}^3 \text{ gold} \times \frac{19.3 \text{ g gold}}{1 \text{ cm}^3 \text{ gold}} = 660 \text{ g gold} \left(\begin{array}{c} \text{rounded to 2} \\ \text{significant figures} \end{array}\right)$$

That's a lot of gold! (1 pound is 454 g, so this is almost a pound and a half!)

***Is the Answer Reasonable?*** At first glance, the answer we obtained seems rather large, considering we're dealing with a very thin coating of gold. Therefore, we need to do a rough check of the arithmetic to see whether the answers are of the right size.

In the first calculation we have the arithmetic

$$15 \times 144 \times 6.45$$

Let's approximate this as

$$15 \times 150 \times 6$$

The product $150 \times 6$ is the same as $150 \times 2 \times 3$; and $150 \times 2 = 300$, so $(150 \times 2) \times 3$ equals 900. The arithmetic, then, is $15 \times (150 \times 6) = 15 \times 900 = 13500$, or $1.35 \times 10^4$. The answer we obtained for the area, $1.4 \times 10^4$ cm$^2$, seems to be okay.

The arithmetic for the thickness is

$$25 \times \frac{10^{-6}}{10^{-2}} = 25 \times 10^{-4}, \text{ or } 2.5 \times 10^{-3}$$

This value agrees with our calculation.

To calculate the volume, the arithmetic is

$$(1.4 \times 10^4) \times (2.5 \times 10^{-3})$$

which we can write as $1.4 \times 2.5 \times 10^4 \times 10^{-3}$. For simplicity, let's round 2.5 to 2; then $1.4 \times 2 = 2.8$, so $1.4 \times 2.5$ will be a bit larger than 2.8 (3.4 makes sense). The product $10^4 \times 10^{-3} = 10^1$. Therefore, $3.4 \times 10^1 = 34$; the answer we obtained for the volume (34 cm$^3$) is reasonable.

The density of gold is nearly 20 g per cubic centimeter, so 34 cm$^3$ of gold would have a mass of about 680 g. Our answer of 660 g seems to be correct.

Doing this approximate arithmetic may seem like an awful lot of work, but it really isn't that bad. We've explained it in great detail here so you can see how the thinking goes. To actually do the arithmetic takes less time than to tell about it.

Below are additional problems dealing with unit conversions that you can work on to practice your problem solving abilities.

**Additional Problems**

1. Iridium (chemical symbol Ir) is a dense, corrosion-resistant metal used to make electrical contacts. What is the mass of a rectangular iridium strip that is 0.10 mm thick, 5.00 mm long, and 2.50 mm wide? The density of iridium is 22.7 g/cm$^3$.

2. The density of water is approximately 1.00 g/mL. What is the weight of 1.00 quart of water in pounds? (1 lb = 453.6 g, 1.0567 qt = 1 L)

3. A submarine's overall density can be lowered and raised by pumping water or air into steel ballast tanks. When the submarine's density is made equal to the density of the surrounding seawater, the submarine will maintain depth, neither rising nor sinking. If a 103,200 kg submarine takes on 2100 kg of water to maintain depth at 1000 feet, where the density of seawater is approximately 1033 kg/m$^3$, what is the total displacement (volume) of the submarine?

4. A rectangular salt water aquarium that is 200.0 cm long and 50.0 cm wide is placed on a table. If the aquarium is filled to a depth of 100.0 cm, how many kilograms of seawater will the table have to support? Assume that the seawater has a density of 1.02 g/cm$^3$.

5. The Committee on Oil Pollution Act of 1990 has barred single hulled tankers from U.S. waters by the year 2010 if they carry more than $5.00 \times 10^3$ tons of oil. If the oil has a density of 0.90 g/mL, how many gallons of oil may a single hulled tanker legally carry? (1 ton = 2000 lb; 453.6 g = 1 lb; 946 mL = 1 qt; 4 qt = 1 gal)

6. The density of jet fuel is about 0.800 g/mL at 0 °C and 0.775 g/mL at 35 °C. If the jet's tank holds a volume of 10,375 L, how many more kg of fuel can the plane carry at 0 °C then at 35 °C?

7. The density of ice is about 0.90 g/mL; the density of water is about 1.00 g/mL. Write a conversion factor that converts mL of ice into mL of water.

**Answers**     **1.** 0.028 g, **2.** 2.09 lb, **3.** 101.9 m³, **4.** $1.02 \times 10^3$ kg, **5.** $1.3 \times 10^6$ gal, **6.** The plane can carry 260 kg more fuel at 0 °C than at 35 °C, **7.** $\dfrac{1.1 \text{ mL ice}}{1 \text{ mL water}}$

# 3. THE MOLE CONCEPT AND STOICHIOMETRIC CALCULATIONS

## The Mole Concept

The most important chemical concept you will encounter during your general chemistry course is that of the mole. It is used in all reasoning that deals with amounts of chemicals involved in compounds and reactions. Because it is at the heart of so much of what we do in chemistry, it is essential that you become comfortable with mole reasoning. If you achieve this sense of "comfort," you are likely to find that the rest of chemistry falls into place relatively easily. On the other hand, if you are forever struggling with the mole concept, much of the rest of chemistry will be a mystery.

*The **mole** is a word that stands for a fixed number of things*, just as a dozen stands for a fixed number. As it happens, the number of things that make up a mole is enormous ($6.02 \times 10^{23}$), but in most situations, we won't actually need to use this number[10], just the notion that when we use the word mole, we mean a certain fixed number of things.

To illustrate how we use the mole concept, let's consider the substance propane, the gas used for cooking in rural areas and in nearly all gas barbecue grills. Propane consists of molecules that have the formula $C_3H_8$.

 A molecule of propane, $C_3H_8$. It consists of three carbon atoms and eight hydrogen atoms joined together in a single particle that we call a molecule.

If we had one molecule of propane, it would contain 3 atoms of C and 8 atoms of H. If we had 10 propane molecules, all together they would contain 30 atoms of C and 80 atoms of H. Notice, however, that even though we have more atoms of C and H in the larger sample, they are in the same numerical C-to-H ratio, namely 3-to-8. In fact, in any sample of this compound, the atom *ratio* of C to H will be 3 to 8.

Now suppose we had one dozen propane molecules. In this sample we would find 3 dozen carbon atoms and 8 dozen hydrogen atoms[11].

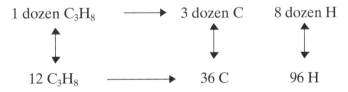

---

[10]When the mole concept was first "invented" the number of things in a mole was not able to be measured, but this did not prevent the concept from being used.

[11]When we write "1 dozen $C_3H_8$" we mean "1 dozen $C_3H_8$ *molecules*," and when we write "3 dozen C" and "8 dozen H," we mean "3 dozen C *atoms*" and "8 dozen H *atoms*." We generally omit the word molecule or atom because the formula $C_3H_8$ stands for the molecule of $C_3H_8$, and the symbols C and H stand for atoms of C and H.

Notice that in a dozen molecules, the ratio of atoms *by the dozen* (3 to 8) is numerically the same as the atom ratio in one molecule of the compound.

The mole (symbol, **mol**) is a quantity similar to the dozen, because it stands for a fixed number of things. The reasoning with moles is therefore the same as the reasoning with dozens. If we had one mole of $C_3H_8$, the ratio of atoms *by the mole* would be numerically the same as the atom ratio in one molecule.

$$1 \text{ mol } C_3H_8 \rightarrow 3 \text{ mol C} \qquad 8 \text{ mol H}$$
$$1 \text{ dozen } C_3H_8 \rightarrow 3 \text{ dozen C} \qquad 8 \text{ dozen H}$$
$$1 \text{ molecule } C_3H_8 \rightarrow 3 \text{ atom C} \qquad 8 \text{ atom H}$$

Thus, whether we're dealing with individual particles (molecules and atoms), particles by the dozen, or particles by the mole, the ratios are the same.

### Relating Moles of One Substance to Moles of Another in a Compound

**Skills Analysis**

1. Submarine life support systems can generate oxygen gas ($O_2$) by passing a strong electric current through water, $H_2O$. For every 1000 molecules of oxygen gas produced, how many molecules of water are consumed?

2. An electric spark can convert oxygen gas ($O_2$) into ozone ($O_3$). Write a conversion factor that converts moles of $O_2$ into moles of $O_3$.

3. The smell of the sea comes from a gas called dimethyl sulfide, $(CH_3)_2S$. How many moles of H are in 0.0230 moles of dimethyl sulfide?

**Answers**

If you missed any of these questions, read this entire section.

1. 2000 molecules $H_2O$, 2. $\dfrac{2 \text{ mol } O_3}{3 \text{ mol } O_2}$, 3. 0.138 mol H

**Review**

In chemical problem solving, the chemical formula serves as a tool that establishes the ratios among atoms in a single molecule of the compound. It is also a tool that establishes the ratios among moles of atoms in a mole of the compound. We can use these relationships to construct equivalencies that we can use in chemical calculations. For propane, we can write the following, which apply when we are dealing with numbers of individual molecules and atoms.

$$1 \text{ molecule } C_3H_8 \Leftrightarrow 3 \text{ atom C}$$
$$1 \text{ molecule } C_3H_8 \Leftrightarrow 8 \text{ atom H}$$
$$3 \text{ atom C} \Leftrightarrow 8 \text{ atom H}$$

For example, the first equivalence would be used when we are interested in relating the number of carbon atoms to a certain number of $C_3H_8$ molecules. It tells us that for every 1 molecule of $C_3H_8$, we will find 3 atoms of C. If we were interested in the relation between molecules of $C_3H_8$

and atoms of H, then we would use the second equivalence, and if we wanted to relate atoms of C and atoms of H in a sample of propane, we would use the third equivalence.

Individual atoms and molecules are too small to work with in the laboratory, so we scale everything up to mole-sized quantities and write the following equivalencies.

$$1 \text{ mol } C_3H_8 \Leftrightarrow 3 \text{ mol } C$$
$$1 \text{ mol } C_3H_8 \Leftrightarrow 8 \text{ mol } H$$
$$3 \text{ mol } C \Leftrightarrow 8 \text{ mol } H$$

Notice that the numbers involved in these equivalencies are identical to those in the relationships given for the individual molecules and atoms; in both cases they are derived from the subscripts in the formula for propane.

Let's look at two examples that illustrate how we use the concepts described above to solve problems.

### PROBLEM 9

How many atoms of hydrogen are found in a sample of butane, $C_4H_{10}$, the fuel used in cigarette lighters, if the sample contains a total of 16 atoms of carbon?

*Analysis:* From the preceding discussion, you probably see where we're going with this, but let's suppose we came across this question "out of the blue." How do we proceed? We can begin by stating the problem in equation format:

$$16 \text{ atom } C \Leftrightarrow \text{? atom } H$$

We need to relate atoms of carbon and atoms of H found in molecules of butane. The critical link is the formula of butane. The set of subscripts in $C_4H_{10}$ is the tool that tells us that in one molecule of $C_4H_{10}$ there are 4 atoms of carbon and 10 atoms of hydrogen. We can write three equivalencies from this information, but we're only interested in the one that relates atoms of C to atoms of H.

$$4 \text{ atom } C \Leftrightarrow 10 \text{ atom } H$$

We can now use this to form a conversion factor to find the answer to the question.

*Solution:* The solution to the problem is

$$16 \text{ atom } C \times \frac{10 \text{ atom } H}{4 \text{ atom } C} \Leftrightarrow 40 \text{ atom } H$$

***Is the Answer Reasonable?*** We can use simple proportional reasoning here to confirm that we've got the correct answer. In 1 molecule of $C_4H_{10}$ there are 4 atom C, so to have 16 atom of C we must have 4 molecules of $C_4H_{10}$. In one molecule of $C_4H_{10}$ there are 10 atoms of H, so in 4 molecules of $C_4H_{10}$ there must be 40 atoms of H.

### PROBLEM 10

When butane burns in air, the carbon in the compound becomes incorporated entirely in molecules of carbon dioxide, $CO_2$. How many moles of $CO_2$ would be formed if 0.200 mol of $C_4H_{10}$ is burned?

***Analysis:*** Let's begin by expressing the problem in an equation format:

$$0.200 \text{ mol } C_4H_{10} \Leftrightarrow \text{ ? mol } CO_2$$

The critical link in solving this problem is finding the amount of carbon (in moles) in the sample of $C_4H_{10}$. Because all the carbon in the $C_4H_{10}$ goes into the $CO_2$, we can then use this amount of carbon to figure out how much $CO_2$ is formed. The tools we will use are the subscripts in the formulas of $C_4H_{10}$ and $CO_2$. Here's an outline of the path we will use to solve the problem.

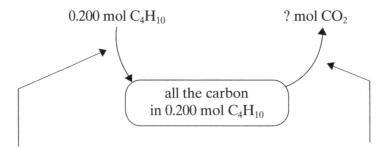

0.200 mol $C_4H_{10}$           ? mol $CO_2$

all the carbon in 0.200 mol $C_4H_{10}$

<u>Tool</u>: Subscript of C in $C_4H_{10}$           <u>Tool</u>: Subscript of C in $CO_2$
4 mol C $\Leftrightarrow$ 1 mol $C_4H_{10}$           1 mol C $\Leftrightarrow$ 1 mol $CO_2$

***Solution:*** We use the tools to form conversion factors. As usual, we are careful to select the factors that allow the units to cancel.

$$0.200 \text{ mol } C_4H_{10} \times \frac{4 \text{ mol C}}{1 \text{ mol } C_4H_{10}} \times \frac{1 \text{ mol } CO_2}{1 \text{ mol C}} = 0.800 \text{ mol } CO_2$$

***Is the Answer Reasonable?*** Let's do some whole-number reasoning to see how the relative amounts of $C_4H_{10}$ and $CO_2$ compare. If we had 1 mol $C_4H_{10}$, it would contain 4 mol C. But it takes only 1 mol of C to make 1 mol of $CO_2$, so 4 mol of C is enough to make 4 mol of $CO_2$. The conclusion, then, is that 1 mol of $C_4H_{10}$ has enough carbon in it to make 4 mol of $CO_2$. Another way of looking at this is that the number of moles of $CO_2$ formed is four times the number of moles of $C_4H_{10}$. If we start with 0.200 mol $C_4H_{10}$, then the amount of $CO_2$ formed would be $4 \times 0.200 = 0.800$ mol. That's the answer we obtained.

1. A compound containing only carbon and sulfur was completely burned in a sealed container. The container held 0.17 mol of $CO_2$ and 0.34 mol of $SO_2$ after the compound burned completely. What was the formula of the compound?

2. Chlorine atoms cause chain reactions in the stratosphere that destroy the Earth's ozone layer. The chlorine atoms come from chlorofluoro-carbons like $CF_2Cl_2$, which were used for many years as refrigerants. $CF_2Cl_2$ breaks down in a complex series of reactions with light and oxygen to form Cl atoms and fluorine-containing molecules like $COF_2$. If a sample contains 35 $COF_2$ molecules, how many Cl atoms should be found if $CF_2Cl_2$ is the only source of F and Cl atoms in the sample?

3. The mineral pyrite ($FeS_2$) occurs naturally in coal beds. When the coal is mined, the pyrite is exposed to air and rain. A sequence of chemical reactions transforms $FeS_2$ into sulfuric acid, $H_2SO_4$. The sulfuric acid in runoff can devastate aquatic ecosystems around the mining opera-tion. If every ton of soft coal mined ultimately results in the decompo-sition of 11 moles of $FeS_2$, how many moles of $H_2SO_4$ could be released when 2.0 million tons of coal are mined, if all of the sulfur in the $H_2SO_4$ comes from $FeS_2$?

**Additional Problems**

1. $CS_2$, 2. 70 atoms of Cl, 3. $4.4 \times 10^7$ mol $H_2SO_4$

**Answers**

## Relating Mass and Moles

1. How many grams of copper are in 0.00100 mol of copper?
2. How many moles of Cu are in 2.5 g of Cu?
3. How many grams of Cl are contained in a sample $CF_2Cl_2$ for each 1.00 gram of F in the sample?
4. What is the mass, in grams, of 0.250 mol of $CF_2Cl_2$?
5. How many moles of $H_2$ could be obtained from the electrolysis of 100.0 g of $H_2O$?
6. How many grams of Al would be needed to make exactly 1 mole of alum [$KAl(SO_4)_2 \cdot 12H_2O$]?

**Skills Analysis**

If you were able to answer Questions 1–3 correctly, you may skip "Atomic Mass as a Tool in Calculating Moles of an Element." You may skip this entire section if you answered Questions 4–6 correctly.
1. 0.0635 g Cu, 2. 0.039 mol Cu, 3. 1.87 g Cl, 4. 30.2 g $CF_2Cl_2$, 5. 5.551 mol $H_2$, 6. 26.98 g Al

**Answers**

To be able to use the mole concept in a practical way, we need a method to measure moles of things in the laboratory. The foundation for these measurements is the fact that each element has its own characteristic

**Atomic Mass as a Tool in Calculating Moles of an Element**

atomic mass[12], which differs from the atomic masses of other elements. Carbon, for example, has an atomic mass of 12.01, whereas that of magnesium is 24.30, which tells us that an atom of magnesium is about twice as heavy as one of carbon. Therefore, if we had a sample of magnesium with the same number of atoms as a sample of carbon, the magnesium sample would weight about twice as much. This notion forms the basis for a way of defining how large a mole is: *One mole of an element has enough atoms so that its mass measured in grams is numerically equal to the atomic mass of the element.* Thus, one mole of carbon atoms (1 mol C) has a mass of 12.01 g, and one mole of magnesium atoms (1 mol Mg) has a mass of 24.30 g.

The point of the preceding discussion is that a table of atomic masses (or a periodic table that incorporates atomic masses) provides us with a tool for relating moles of elements to masses measured in grams. All we need to do is look up the atomic mass of the element and we can write down how much one mole of that element weighs. For example, the periodic table gives the atomic mass of sodium (Na) as 22.98977, so we can write

$$1 \text{ mol Na} = 22.98977 \text{ g Na}$$

Usually, this is more significant figures than we require, so we round it as needed. If four significant figures are sufficient, then we write

$$1 \text{ mol Na} = 22.99 \text{ g Na}$$

Here are two more examples, also to four significant figures.

$$1 \text{ mol Cl} = 35.45 \text{ g Cl}$$

$$1 \text{ mol Fe} = 55.85 \text{ g Fe}$$

Let's look at two sample problems that require us to relate grams to moles.

### PROBLEM 11

Suppose we needed to measure 2.50 mol of sulfur for a particular experiment. How many grams of sulfur would we need?

*Analysis:* We'll begin by stating the problem in equation form.

$$2.50 \text{ mol S} = ? \text{ g S}$$

To solve the problem, we need to relate moles of sulfur to grams of sulfur. The tool to accomplish this is the atomic mass of sulfur. In the periodic table, this is given as 32.066. To be sure to have enough preci-

---

[12]As noted in the text, naturally occurring samples of almost all elements consist of mixtures of two or more isotopes that have slightly different masses. The atomic masses given in the periodic table correspond to the average atomic mass of the atoms in the mixture and these average masses are essentially the same regardless of the origin of the sample. It's convenient, therefore, to think of a reported atomic mass as the mass of an average atom of the element.

sion in the calculation, we'll round this to one more significant figure than in the given data and write

$$1 \text{ mol S} = 32.07 \text{ g S}$$

Now we can use this to form a conversion factor to solve the problem.

***Solution:*** We form the conversion factor so mol S will cancel.

$$2.50 \text{ mol S} \times \frac{32.07 \text{ g S}}{1 \text{ mol S}} = 80.2 \text{ g S}$$

***Is the Answer Reasonable?*** Some simple arithmetic will answer the question. One mole of sulfur is about 32 g. We have between two and three moles of S, so the answer should lie between $2 \times 32 \text{ g} = 64 \text{ g}$ and $3 \times 32 \text{ g} = 96 \text{ g}$. Our answer, 80.2 g, is between these two values, so the answer does seem reasonable.

## PROBLEM 12

A useful cleaning agent sold in many hardware stores is called TSP, which stands for "trisodium phosphate." The formula of the compound is $Na_3PO_4$. Suppose a sample of this substance contains 16.4 g of Na. How many grams of phosphorus does it contain?

***Analysis:*** Once again, let's begin by stating the problem in equation form.

$$16.4 \text{ g Na} \Leftrightarrow ? \text{ g P}$$

To find the solution, we need to ask ourselves "What tools do need?" How can we relate the amount of Na to the amount of P in this compound?

To answer this question we need to recognize that the critical link is the chemical formula. It relates amounts of Na and P, but the relationship is in moles, not grams. Thus, from $Na_3PO_4$, we can write

$$3 \text{ mol Na} \Leftrightarrow 1 \text{ mol P}$$

To complete the flow of units, however, we need to relate grams of Na to moles of Na, and moles of P to grams of P.

$$16.4 \text{ g Na} \Leftrightarrow ? \text{ g P}$$

$$3 \text{ mol Na} \Leftrightarrow 1 \text{ mol P}$$

These tools are obtained from the atomic masses. Rounding them to four significant figures, we can write

$$1 \text{ mol Na} = 22.99 \text{ g Na}$$

$$1 \text{ mol P} = 30.97 \text{ g P}$$

Now we have all the information we need to go from g Na to g P.

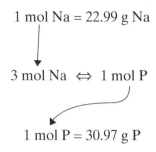

1 mol Na = 22.99 g Na

3 mol Na ⟺ 1 mol P

1 mol P = 30.97 g P

***Solution:*** We now set up the solution by forming conversion factors so the units cancel (add the cancel marks yourself.)

$$16.4 \text{ g Na} \times \frac{1 \text{ mol Na}}{22.99 \text{ g Na}} \times \frac{1 \text{ mol P}}{3 \text{ mol Na}} \times \frac{30.97 \text{ g P}}{1 \text{ mol P}} = 7.36 \text{ g P}$$

The answer is that a sample that contains 16.4 g Na also contains 7.36 g P.

***Is the Answer Reasonable?*** Let's do some proportional reasoning here along with some approximate arithmetic. On a mole basis, the formula $Na_3PO_4$ tells us that the amount of Na present is three times the amount of phosphorus. Another way of looking at this is that the number of moles of phosphorus is one third as large as the number of moles of sodium. If the atomic masses of Na and P were the same, then the mass of P would be one third the mass of Na. If we approximate the mass of Na as 15 g, then the amount of P would be 5 g. But the actual mass of Na (16.4 g) is larger than 15 g, and the actual atomic mass of P is larger than that of Na, so the actual mass of P should be somewhat larger than 5 g. Our answer of 7.63 is a bit larger than 5 g, so the answer does seem reasonable.

In our reasoning here, we've made some pretty big approximations, but the goal was to check that the answer we obtained is not grossly in error. If we had used any of the conversion factors upside down, the answer would have been far from 5 g and we would have known that a major mistake had been made.

**The Chemical Formula as a Tool Used to Calculate Molecular or Formula Masses**

When we add up the atomic masses of all the atoms in a molecule we obtain the **molecular mass** of the compound. Thus, the molecular mass of propane, $C_3H_8$, is computed as follows:

$$
\begin{array}{lll}
\text{carbon:} & 3 \times 12.01 = & 36.03 \\
\text{hydrogen:} & 8 \times 1.01 \ \ = & \underline{8.08} \\
& \text{Total} = & 44.11
\end{array}
$$

In Chapter 2 you learned that ionic compounds such as NaCl do not contain discrete molecules. Instead, the ions are packed around each other in a way that places ions of opposite charge as close as possible to each other. For ionic compounds, we use the term **formula unit** to describe the set of ions specified by the chemical formula. For example, the formula unit for NaCl contains one $Na^+$ and one $Cl^-$ ion. Similarly, a formula unit of aluminum chloride, $AlCl_3$, consists of one $Al^{3+}$ ion and three

Cl⁻ ions. When we add up the atomic mass of the atoms that make up one formula unit of the compound, we obtain the **formula mass**. For example, the formula mass of aluminum chloride is calculated as

$$
\begin{array}{lrl}
\text{aluminum:} & 1 \times 26.98 = & 26.98 \\
\text{chlorine:} & \underline{3 \times 35.45 = 106.4} \\
& \text{Total} = & 133.4
\end{array}
$$

Any time we have a chemical formula, we can calculate a formula mass. For compounds that actually consist of molecules, such as $C_3H_8$, the formula mass and molecular mass mean the same thing. For ionic compounds, which contain no molecules, only the term formula mass is truly applicable.

---

The reason formula and molecular masses are important is because they tell us the mass of one mole of a compound. For example, a single formula unit of sodium chloride (NaCl) has a formula mass of 58.44, which is the sum of the atomic masses of sodium and chlorine. A mole of NaCl contains one mole of sodium (22.99 g Na) and one mole of chlorine (35.45 g Cl), so the sum of these masses, 58.44 g, is the mass of *one mole* of NaCl. Thus, *the mass in grams of a mole of a compound is numerically the same as the compound's formula mass.* To determine the mass of a mole of a compound, we calculate the formula mass (or molecular mass) and add the unit "gram." Let's look at another example.

**Formula and Molecular Masses as Tools Used in Converting Between Moles and Grams of Substances**

Propane has a molecular mass of 44.09, so a mole of propane has a mass of 44.09 g. We can express this by the equation

$$1 \text{ mol } C_3H_8 = 44.09 \text{ g } C_3H_8$$

Notice that in writing this relationship, we have included the formula $C_3H_8$ along with the units "mol" and "g." It's always wise to include this information so we keep in mind what it is we have moles or grams of.

### PROBLEM 13

What is the mass in grams of 0.650 mol of $Na_3PO_4$?

*Analysis:* We begin by expressing the problem in equation form.

$$0.650 \text{ mol } Na_3PO_4 = ? \text{ g } Na_3PO_4$$

The critical link in solving this problem is realizing that whenever we want to relate mass and moles of a compound, the tool is the formula (or molecular) mass. To solve this problem, therefore, we need to first calculate the formula mass of $Na_3PO_4$.

$$
\begin{array}{llrl}
\text{sodium:} & 3 \times 22.99 = & 68.97 \\
\text{phosphorus:} & 1 \times 30.97 = & 30.97 \\
\text{oxygen:} & \underline{4 \times 16.00 = \ \ 64.00} \\
& \text{Total} = & 163.94
\end{array}
$$

The formula mass of $Na_3PO_4$ is 163.94, so we can write

$$1 \text{ mol } Na_3PO_4 = 163.94 \text{ g } Na_3PO_4$$

We now have the information necessary to solve the problem.

*Solution:* We use the relationship above to form a conversion factor to convert moles to grams.

$$0.650 \text{ mol Na}_3\text{PO}_4 \times \frac{163.94 \text{ g Na}_3\text{PO}_4}{1 \text{ mol Na}_3\text{PO}_4} = 107 \text{ g Na}_3\text{PO}_4$$

The sample has a mass of 107 g.

*Is the Answer Reasonable?* We have slightly more than 0.6 mol of $\text{Na}_3\text{PO}_4$. One tenth of a mole of the compound would weigh about 16 g (actually, 16.394 g), so 0.6 mol would weigh about $6 \times 16 = 94$ g. Our sample, because it is larger than 0.6 mol, should weigh somewhat more than this, so 107 g seems reasonable.

## PROBLEM 14

How many moles of sodium are found in 34.7 g of trisodium phosphate, $\text{Na}_3\text{PO}_4$?

*Analysis:* As usual, we start by expressing the problem as an equation.

$$34.7 \text{ g Na}_3\text{PO}_4 \Leftrightarrow ? \text{ mol Na}$$

Next, we need to select the tools that will allow us to solve the problem. In looking over the problem, we see that we're being asked, in effect, "How much Na is in a sample of $\text{Na}_3\text{PO}_4$?" From our earlier discussions, we know that the chemical formula is a tool that will let us relate moles of Na to moles of $\text{Na}_3\text{PO}_4$, so let's begin with that. From the formula, we can write

$$1 \text{ mol Na}_3\text{PO}_4 \Leftrightarrow 3 \text{ mol Na}$$

This relationship contains the desired unit "mol Na", so now we need to relate the given mass of $\text{Na}_3\text{PO}_4$ to moles of $\text{Na}_3\text{PO}_4$. The tool to accomplish this is the formula mass of $\text{Na}_3\text{PO}_4$. In the preceding problem, we calculated the formula mass to be 163.94, so

$$1 \text{ mol Na}_3\text{PO}_4 = 163.94 \text{ g Na}_3\text{PO}_4$$

We now have the information we need to go from "g $\text{Na}_3\text{PO}_4$" → "mol $\text{Na}_3\text{PO}_4$" → "mol Na." Follow the path through the units.

$$1 \text{ mol Na}_3\text{PO}_4 \quad = \quad 163.94 \text{ g Na}_3\text{PO}_4$$

$$1 \text{ mol Na}_3\text{PO}_4 \quad \Leftrightarrow \quad 3 \text{ mol Na}$$

*Solution:* We use the tools we selected to form conversion factors. As usual, we make sure the units cancel correctly. You can add the cancel marks yourself.

$$34.7 \text{ g Na}_3\text{PO}_4 \times \frac{1 \text{ mol Na}_3\text{PO}_4}{163.94 \text{ g Na}_3\text{PO}_4} \times \frac{3 \text{ mol Na}}{1 \text{ mol Na}_3\text{PO}_4} = 0.635 \text{ mol Na}$$

***Is the Answer Reasonable?*** A tenth of a mole of $Na_3PO_4$ has a mass of about 16 g. Our sample is 34 g, which is about $2 \times 16$, so we have approximately 0.2 mol of $Na_3PO_4$. In 1 mol $Na_3PO_4$ there are 3 mol Na, so in 0.2 mol of $Na_3PO_4$ there would be 0.6 mol of Na. Our answer of 0.635 mol is close to this, so our approximate arithmetic suggests that the answer is reasonable.

---

**A Look Back**

Before moving on, it's worth reviewing what was involved in solving the preceding problem. First, the critical link in finding the solution was realizing that the chemical formula allows us to relate the amount of Na in the $Na_3PO_4$. Without that understanding, we would not have been able to proceed. Second, the solution to the problem requires two steps and the application of two of the chemical tools we've discussed so far in this Supplement. One is the mole relationship between Na and $Na_3PO_4$ provided by the chemical formula; the other is the formula mass of $Na_3PO_4$, which relates grams of $Na_3PO_4$ to moles of $Na_3PO_4$. These are tools that are summarized in the table on Page 139 of the textbook.

---

**Additional Problems**

1. How many grams of plutonium are in 1.00 mmol of plutonium?
2. How many moles of osmium are in 1.00 g of osmium?
3. How many grams of $O_2$ are required to completely convert 120.1 g of C to $CO_2$?
4. How many grams of CuO can be made from a pure copper wire with a mass of 5.0 g?
5. What is the mass, in g, of 0.250 mol of phenobarbital ($C_{12}H_{12}N_2O_3$)?
6. How many grams of Cl are contained in 0.100 mol of the drug chloral hydrate ($C_2H_3Cl_3O_2$)?
7. How many moles of pentachlorophenol ($C_6HCl_5O$) contain 1.00 g of Cl?

**Answers**

1. 0.244 g Pu, 2. 0.00526 mol Os, 3. 320.0 g $O_2$, 4. 6.3 g CuO, 5. 58.1 g $C_{12}H_{12}N_2O_3$, 6. 10.6 g Cl, 7. 0.00564 mol $C_6HCl_5O$

## Relating Masses of Substances in Compounds

**Skills Analysis**

1. If 40.0 g of a metal reacts with oxygen to form 56.0 g of metal oxide, what is the percentage of metal in the metal oxide?
2. Write a conversion factor that converts grams of chalk (as g $CaCO_3$) into grams of calcium contained in the chalk.
3. What is the percentage of barium by mass in the mineral barite ($BaSO_4$)?
4. How many grams of nitrogen are in 1.00 kg of the fertilizer $NH_4NO_3$?
5. How many grams of gold(III) chloride ($AuCl_3$) can be made from 1.00 g of gold?

**Answers**

**1.** 71.4% metal by mass, **2.** $\dfrac{40.078 \text{ g Ca}}{100.087 \text{ g CaCO}_3}$, **3.** 58.8400% Ba by mass, **4.** 350 g N, **5.** 1.54 g AuCl$_3$

If you missed questions 1, 2, or 3, you should read this entire section. If you missed questions 4 or 5, you should read "Relationships between Masses of Elements in a Compound."

**Percentages**

In the preceding calculations, we related grams to moles both for elements and for compounds. Sometimes, however, we simply wish to know the relationships among the masses of the components of a compound. One way such mass relationships are expressed is as a percentage by mass.

In general, a percentage is calculated by taking the ratio of "a part to the whole" and then multiplying that ratio by 100 %. For example, suppose a $500 TV was on sale for $450, a price reduction of $50. The price reduction amounts to 10% of the whole regular price.

part of the regular price
that is the price reduction

$$\text{percent price reduction} = \frac{\$50}{\$500} \times 100 \text{ \%} = 10 \text{ \%}$$

whole regular price

**Percentage Composition of Compounds**

The calculation of the percentage by mass of the elements in a compound follows a straightforward procedure. Consider, for example, the compound $Na_2CO_3$. One mole of this substance contains two moles of sodium, one mole of carbon, and three moles of oxygen. We can use these amounts and the atomic masses of the elements to calculate the masses of each in one mole of the compound.

| Element | Moles | Mass |
|---------|-------|------|
| Sodium | 2 mol | $2 \times 22.99$ g = 45.98 g |
| Carbon | 1 mol | $1 \times 12.01$ g = 12.01 g |
| Oxygen | 3 mol | $3 \times 16.00$ g = 48.00 g |
| | | Total mass    105.99 g |

Once we have the masses of each element and the formula mass, we can calculate the percentages by mass of each. Let's start with the percent sodium.

mass of sodium in 1 mol $Na_2CO_3$

$$\text{percent sodium} = \frac{45.88 \text{ g}}{105.99 \text{ g}} \times 100 \text{ \%} = 43.38 \text{ \% Na}$$

mass of 1 mol $Na_2CO_3$

Similarly, for carbon and oxygen,

$$\text{percent carbon} = \frac{12.01 \text{ g}}{105.99 \text{ g}} \times 100\% = 11.33\% \text{ C}$$

$$\text{percent oxygen} = \frac{48.00 \text{ g}}{105.99 \text{ g}} \times 100\% = 45.29\% \text{ O}$$

Sum of percentages = 100.00 %

Calculating a percentage composition doesn't involve a great deal of reasoning. It's just a matter of learning how to do the calculation. It is helpful to remember that the sum of the percentages should add up to 100 % (although sometimes the calculated sum will be slightly more or slightly less than 100 % because of rounding during the calculations).

There are times when we need to be able to calculate the mass of an element in a sample of a compound. For example, suppose we wished to know how many grams of sodium are in 3.58 g of $Na_2CO_3$. One way to do the calculation is the following:

**Relationships Between Masses of Elements in a Compound**

1. Use the formula mass of $Na_2CO_3$ to calculate the moles of $Na_2CO_3$ in the sample.
2. Use the subscripts in $Na_2CO_3$ to calculate the moles of Na in the sample.
3. Use the atomic mass of Na to calculate the grams of Na in the sample.

There is a faster way to solve the problem, however. In the process of calculating the formula mass of $Na_2CO_3$ we obtain all the data we need to do the calculation in one step instead of three. This is because in calculating the formula mass we obtain the data required to establish a mass relationship between Na and $Na_2CO_3$.

| Element | Moles | Mass |
|---------|-------|------|
| Sodium | 2 mol | $2 \times 22.99 \text{ g} = 45.98 \text{ g}$ |
| Carbon | 1 mol | $1 \times 12.01 \text{ g} = 12.01 \text{ g}$ |
| Oxygen | 3 mol | $3 \times 16.00 \text{ g} = 48.00 \text{ g}$ |
| Total mass of 1 mol $Na_2CO_3$ | | 105.99 g |

$$105.99 \text{ g } Na_2CO_3 \Leftrightarrow 45.98 \text{ g Na}$$

We can use this relationship to solve the problem,

$$3.58 \text{ g } Na_2CO_3 \times \frac{45.98 \text{ g Na}}{105.99 \text{ g } Na_2CO_3} = 1.55 \text{ g Na}$$

Let's look at another problem that is a variation of this.

**PROBLEM 15**

What is the maximum mass of chromium(III) oxide ($Cr_2O_3$, a green pigment used in paints) that could be made from 26.4 g of Cr?

*Analysis:* We can begin by expressing the problem in equation form:

$$26.4 \text{ g Cr} \Leftrightarrow ? \text{ g Cr}_2O_3$$

The problem relates the mass of Cr that will be in some final mass of the compound $Cr_2O_3$. Let's calculate the formula mass of the chromium(III) oxide.

| Element | Moles | Mass |
|---------|-------|------|
| Chromium | 2 mol | $2 \times 52.00 \text{ g} = 104.0 \text{ g}$ |
| Oxygen | 3 mol | $3 \times 16.00 \text{ g} = 48.00 \text{ g}$ |
| | | Total      152.0 g |

The data in the table permits us to write the equivalence

$$104.0 \text{ g Cr} \Leftrightarrow 152.0 \text{ g Cr}_2O_3$$

This is precisely the relationship we need to do the calculation.

*Solution:* Assembling and applying the conversion factor gives,

$$26.4 \text{ g Cr} \times \frac{152.0 \text{ g Cr}_2O_3}{104.0 \text{ g Cr}} = 38.6 \text{ g Cr}_2O_3$$

The answer tells us that if we incorporate 26.4 g of Cr into the compound $Cr_2O_3$, the mass of the compound will be 38.6 g.

***Is the Answer Reasonable?*** Some simple approximate arithmetic can be done here. First, in the equivalence above we see that there are about 100 g of Cr in about 150 g of $Cr_2O_3$. This tells us that the mass of $Cr_2O_3$ is about 1.5 times larger than the mass of Cr. The amount of Cr we have in the problem is approximately 26 g, and 1.5 times this is 39 g. Therefore, our answer of 38.6 g does seem reasonable.

---

**Additional Problems**

1. Pure titanium is prepared by heating $TiCl_4$ with sodium in a steel "bomb." How many grams of Ti can be obtained from 100.0 g of $TiCl_4$?

2. People exposed to fallout from nuclear weapons are given potassium iodide (KI) tablets. The iodine in the tablets floods the thyroid gland and prevents radioactive iodine in the fallout from binding there. A tablet containing 1.50 g of KI supplies how many grams of iodine?

3. Which of these minerals has the highest percentage of copper by mass: bornite ($Cu_5FeS_4$), cuprite ($Cu_2O$), or chalcopyrite ($CuFeS_2$)?

4. What is the percentage of mercury by mass in the contact lens solution ingredient thimerosal ($C_2H_5HgSC_6H_4CO_2Na$)?

5. A metal alloy can be analyzed for aluminum by dissolving the metal in acid and then treating the solution to precipitate the aluminum as $Al(C_9H_6ON)_3$. If 2.000 g of $Al(C_9H_6ON)_3$ is obtained, how many grams of aluminum were in the original sample?

**1.** 25.24 g Ti, **2.** 1.15 g I, **3.** Cuprite (88.8% Cu), **4.** 49.55% Hg, **5.** 0.1175 g Al   **Answers**

### Dealing with Individual Atoms and Molecules

1. How many atoms are there in 0.5 moles of iron?

2. 25.0 moles of phosphoric acid ($H_3PO_4$) contain how many hydrogen atoms?

3. What is the mass of one atom of plutonium, in grams?

4. How many atoms of uranium are there in 1.00 g of uranium metal?

5. How many $H_2O$ molecules are there in a raindrop weighing 0.050 g?

**Skills Analysis**

If you miss any of the questions on the pretest, you should read this entire section.   **Answers**

**1.** $3 \times 10^{23}$ atoms Fe, **2.** $4.52 \times 10^{25}$ atoms H, **3.** $4.05 \times 10^{-22}$ g Pu,

**4.** $2.53 \times 10^{21}$ atoms U, **5.** $1.7 \times 10^{21}$ molecules $H_2O$

As you've learned, the mole is a unit that stands for an fixed number of things. This number is called Avogadro's number and is equal to $6.02 \times 10^{23}$.

$$1 \text{ mol} = 6.02 \times 10^{23} \text{ things}$$

**The Mole and Avogadro's Number**

For most purposes, it isn't necessary to know the size of this number. Usually, it is sufficient to understand that if we have equal numbers of moles of two elements, they contain equal numbers of atoms. However, there are some instances when we do need to know the number of things in a mole, and we will study those now.

To understand when we need to use Avogadro's number, we need to examine how we view chemical substances from two perspectives—the large scale *macroscopic* world of the laboratory (where we deal with physical samples of substances and describe amounts in grams and moles) and the very tiny *submicroscopic* world of atoms and molecules (where we count individual atoms and molecules). As long as we stay entirely within one of these two views of matter, we don't need Avogadro's number. Thus, to calculate the number of *moles* of carbon in 5 *moles* of glucose, $C_6H_{12}O_6$, Avogadro's number isn't necessary. We simply use the chemical formula to tell us how many moles of carbon are in one mole of $C_6H_{12}O_6$. Similarly, if we wished to know how many *atoms* of carbon are in 5 *molecules* of glucose, $C_6H_{12}O_6$, we use the chemical formula to tell us how many carbon atoms are in one molecule of $C_6H_{12}O_6$. Once again, we don't need Avogadro's number.

Avogadro's number becomes necessary when we wish to relate an amount of something in the macroscopic world to an amount in the submicroscopic world. An example would be finding the mass in *grams* (a unit in the macroscopic world) of some number of carbon *atoms* (a unit in the submicroscopic world).

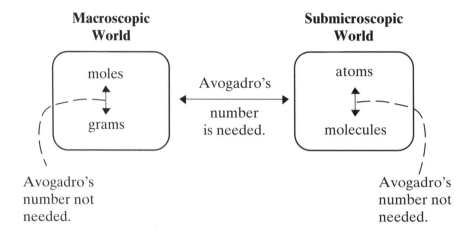

Besides understanding *when* you need to use Avogadro's number, you also need to know *how* to use it. As a tool in calculations, Avogadro's number provides a conversion between moles and individual units of a substance. For example, if we are dealing with an element such as sodium, we can write

$$1 \text{ mol Na} = 6.02 \times 10^{23} \text{ atom Na}$$

For the compound $C_3H_8$, we can write

$$1 \text{ mol } C_3H_8 = 6.02 \times 10^{23} \text{ molecule } C_3H_8$$

For an ionic compound such as NaCl, which doesn't contain discrete molecules, we write

$$1 \text{ mol NaCl} = 6.02 \times 10^{23} \text{ formula units NaCl}$$

There are two important things to observe in these equations. First, notice that we are relating *moles* in the macroscopic world to atoms, molecules, or formula units in the submicroscopic world. Second, notice that when we write these equations we are very careful to specify, using chemical formulas, the exact nature of the substances involved. Now let's look at some sample problems.

### PROBLEM 16
What is the mass in grams of 25 atoms of nickel?

*Analysis:* We'll begin by expressing the problem in the form of an equation:

$$25 \text{ atom Ni} = ? \text{ g Ni}$$

The critical link in solving this problem is realizing that we are attempting to convert a unit in the submicroscopic world (atoms) to a unit in the macroscopic world (grams). Because we are connecting these two realms, we will need to use Avogadro's number as the tool. We are dealing with an element, so we write

$$1 \text{ mol Ni} = 6.02 \times 10^{23} \text{ atom Ni}$$

We can now go from the unit "mol Ni" to "atom Ni." Next, we need a tool to take us from "mol Ni" to "g Ni." This is provided by the atomic mass of nickel.

$$1 \text{ mol Ni} = 58.7 \text{ g Ni}$$

We now have a path from "atom Ni" to "g Ni," so we can proceed to the solution step.

**Solution:** As usual, we assemble conversion factors from the relationships we've established in a way that will allow us to cancel units correctly. (Add the cancel marks yourself.)

$$25 \text{ atom Ni} \times \frac{1 \text{ mol Ni}}{6.02 \times 10^{23} \text{ atom Ni}} \times \frac{58.7 \text{ g Ni}}{1 \text{ mol Ni}} = 2.44 \times 10^{-21} \text{ g Ni}$$

Thus, 25 atoms of nickel have a mass of $2.44 \times 10^{-21}$ g.

**Is the Answer Reasonable?** In working these kinds of problems, the most common mistake is to use Avogadro's number incorrectly, or to not use it at all. Let's see how you can avoid these pitfalls. We know atoms are very tiny and that individual atoms have very small masses. The answer we obtained is a very small number, so in that sense, it seems that our answer reasonable.

If we had used Avogadro's number incorrectly, the answer would have been a huge number. Atoms do not have huge masses, so we should recognize our mistake. Similarly, if we had not used Avogadro's number at all, our calculation would give an answer that is not an extremely tiny number. That should sound an alarm and suggest that we'd better examine our method more closely.

**PROBLEM 17**
How many atoms of carbon are in 0.0250 g of $C_4H_{10}$?

**Analysis:** We begin by expressing the problem in the form of an equation.

$$0.0250 \text{ g } C_4H_{10} \Leftrightarrow \text{? atom C}$$

Once again, the critical link is recognizing that we are trying to relate a unit in the macroscopic world (*gram* $C_4H_{10}$) to a unit in the submicroscopic world (*atom* C). This means we are going to have to use Avogadro's number in the calculation. Where we use it in the calculation depends, in this case, on when we make the cross-over from the macroscopic to the submicroscopic. Let's examine two paths we could follow to the answer, looking at the units only and not worrying, for now, about the numbers.

Path 1.

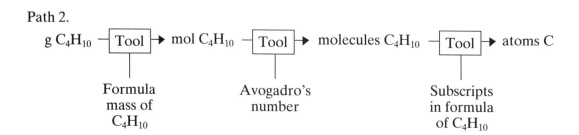

Path 2.

g $C_4H_{10}$ — Tool → mol $C_4H_{10}$ — Tool → molecules $C_4H_{10}$ — Tool → atoms C

Formula
mass of
$C_4H_{10}$

Avogadro's
number

Subscripts
in formula
of $C_4H_{10}$

Notice that in Path 1, we stay at the mole level until the last step. In Path 2, we calculate the number of molecules in the sample, and then calculate how many atoms of carbon are in the $C_4H_{10}$ molecules. Both paths take us to the same place, but they just follow a different course of thinking. Also, notice that we use the same tools in both paths, so the numbers in the calculations will be the same. Let's set up the conversions for each path and then we'll work the solution for each.

Path 1.

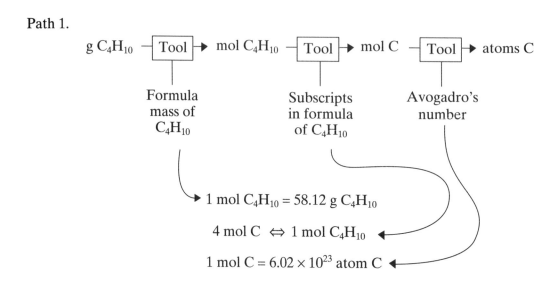

g $C_4H_{10}$ — Tool → mol $C_4H_{10}$ — Tool → mol C — Tool → atoms C

Formula
mass of
$C_4H_{10}$

Subscripts
in formula
of $C_4H_{10}$

Avogadro's
number

1 mol $C_4H_{10}$ = 58.12 g $C_4H_{10}$

4 mol C ⇔ 1 mol $C_4H_{10}$

1 mol C = $6.02 \times 10^{23}$ atom C

Path 2.

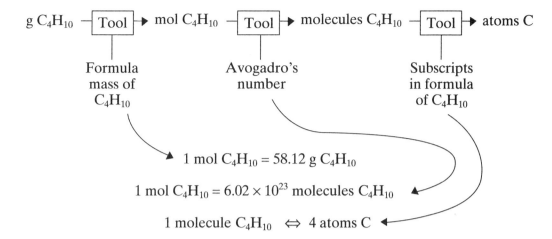

$$1 \text{ mol } C_4H_{10} = 58.12 \text{ g } C_4H_{10}$$

$$1 \text{ mol } C_4H_{10} = 6.02 \times 10^{23} \text{ molecules } C_4H_{10}$$

$$1 \text{ molecule } C_4H_{10} \Leftrightarrow 4 \text{ atoms C}$$

*Solution:* We'll set up the solution for Path 1.

$$0.0250 \text{ g } C_4H_{10} \times \frac{1 \text{ mol } C_4H_{10}}{58.12 \text{ g } C_4H_{10}} \times \frac{4 \text{ mol C}}{1 \text{ mol } C_4H_{10}} \times \frac{6.02 \times 10^{23} \text{ atom C}}{1 \text{ mol C}}$$

$$= 1.04 \times 10^{21} \text{ atom C}$$

The calculation tells us that in 0.0250 g $C_4H_{10}$ there are $1.04 \times 10^{21}$ atoms of carbon.

Now, the calculation for Path 2.

$$0.0250 \text{ g } C_4H_{10} \times \frac{1 \text{ mol } C_4H_{10}}{58.12 \text{ g } C_4H_{10}} \times \frac{6.02 \times 10^{23} \text{ molecule } C_4H_{10}}{1 \text{ mol } C_4H_{10}}$$

$$\times \frac{4 \text{ atom C}}{1 \text{ molecule } C_4H_{10}} = 1.04 \times 10^{21} \text{ atom C}$$

The answer, of course, is the same as we obtained following Path 1.

*Is the Answer Reasonable?* Although the sample size is small, it's still an amount we could measure in the lab, and in any lab-sized amount of a substance, there's a huge number of atoms. The answer we obtained is a very large number, so the answer does seem to be reasonable.

The preceding problem illustrates one of the interesting things about problem solving—sometimes there's more than one way to find the solution. Even though we used the same basic tools, the way they were applied along the two paths to the solution were slightly different, reflecting different ways of thinking about the problem. As long as all the pieces of the puzzle fit together and there is a valid path from the units given in the problem to those required in the answer, you will obtain the correct answer.

1. In 1961, nuclear scientists bombarded a 3 μg target made of the element californium (Cf) with boron nuclei, and synthesized a new element, lawrencium. How many atoms of californium were in the target?

2. How many molecules does a teaspoon of cooking oil contain? Assume that a teaspoon of the oil has a mass of about 4.5 g, and the average molecular mass of the oil is 289 g/mol.

3. How many cobalt atoms are there in 1.0 g of vitamin B-12? (Vitamin B-12 is $C_{63}H_{88}N_{14}O_{14}CoP$, with a molecular mass 1355.4.)

4. Polonium (Po) is an extremely toxic metal. The maximum allowable body burden for polonium is 6.8 pg. How many atoms of polonium is this?

**Answers**

1. $7 \times 10^{15}$ atoms Cf, 2. $9.4 \times 10^{21}$ molecules oil, 3. $4.4 \times 10^{20}$ atoms Co, 4. $2.0 \times 10^{10}$ atoms Po

## Empirical and Molecular Formulas

**Skills Analysis**

1. A compound has an empirical formula $CH_2O$. Which of the following could not be its molecular formula? $C_2H_4O_2$, $CH_2O$, $C_3H_5O_3$, $C_6H_{12}O_6$

2. Heating 1.00 g of europium metal (Eu) in pure oxygen produces 1.16 g of europium oxide. What is the empirical formula of europium oxide?

3. What is the empirical formula of an iron oxide that is 70.0% Fe and 30.0% O by mass?

4. 1.000 g of a compound containing only hydrogen and carbon is burned in pure oxygen gas. 3.064 g of $CO_2$ gas are collected. If all of the carbon in the original compound was trapped in the $CO_2$, what was the empirical formula of the compound?

5. Many sugars have the empirical formula $CH_2O$. What is the molecular formula of a sugar that has molecular mass 180 g mol$^{-1}$?

**Answers**

If you missed questions 1 or 5, you should read "Finding Molecular Formulas." If you missed question 2, read "Empirical Formulas from Direct Elemental Analyses." If you missed question 3, read "Empirical Formulas from Percentage Composition Data." If you missed question 4, read "Empirical Formulas from Indirect Analyses."

1. $C_3H_5O_3$, 2. $Eu_2O_3$, 3. $Fe_2O_3$, 4. $C_3H_7$, 5. $C_6H_{12}O_6$

**Review**

An **empirical formula** (also called a **simplest formula**) provides the smallest whole-number ratio of the atoms in the compound. For example, butane has the empirical formula $C_2H_5$, which tells us that in this substance there are five hydrogen atoms for each two carbon atoms. However, molecules of butane are more complex than the empirical formula suggests. Its molecules each contain 4 atoms of carbon and 10 atoms of hydrogen, so the **molecular formula** is $C_4H_{10}$. Notice that the C-to-H ratio provided by the subscripts in the molecular formula is the same as that given by the subscripts in the empirical formula.

Empirical formula $\quad C_2H_5 \quad$ C-to-H ratio $= \dfrac{2}{5}$

Molecular formula $\quad C_4H_{10} \quad$ C-to-H ratio $= \dfrac{4}{10} = \dfrac{2}{5}$

Thus, the molecular formula contains the information found in the empirical formula, plus it specifies the actual number of atoms of each kind in a molecule of the substance.

Sometimes the empirical and molecular formulas are the same. An example is water, whose molecules contain 2 atoms of H and 1 of O. The formula, of course, is $H_2O$. It is an empirical formula because it has the smallest set of whole-number subscripts and it is a molecular formula as well because it specifies the actual number of atoms of each kind in a molecule of water. For ionic compounds, we always write empirical formulas because ionic compounds do not contain discrete molecules. For ionic compounds, the empirical formula is the formula unit described earlier.

Generally, empirical formulas are derived from data derived from chemical analyses. To obtain the formula, we need to know how the number of moles of the different elements compare. For instance, suppose we analyzed a sample of a compound of carbon and hydrogen and found one mole of carbon and four moles of hydrogen. The C-to-H mole ratio is 1 to 4. This could only be true if the C-to-H atom ratio is *also* 1 to 4, which means that the empirical formula must be $CH_4$.

Formula $\qquad$ C-to-H atom ratio $\qquad$ C-to-H mole ratio
$CH_4 \quad \longleftrightarrow \quad$ 1 to 4 $\quad \longleftrightarrow \quad$ 1 to 4

Notice that the reasoning here is just the opposite of that described earlier, where you learned that if we know the formula, we can determine the mole ratio of elements. Here we are using the mole ratio of elements to determine the formula.

**The Critical Link in Empirical Formula Calculations**

As you know, in a laboratory we don't measure moles directly. Instead, we measure grams and then use atomic or formula masses to convert to moles. In an empirical formula calculation, therefore, the critical data required are the masses of the elements in a sample of a compound. Once these masses are known, we can convert them to moles and determine the mole ratio, which gives us the empirical formula. As you will see, however, the data provided in the problem will not necessarily be the masses of the elements. *Understanding that you must use the data given to find the masses of each of the elements in a sample of the compound is the critical link in being able to calculate the empirical formula.* We will illustrate this by several worked examples.

**Empirical Formulas from Direct Elemental Analyses**

In some instances, we can measure directly the masses of the elements in a sample of a compound. This is the simplest kind of empirical formula problem you will encounter.

**PROBLEM 18**

A sample of aluminum weighing 1.45 g was heated in an atmosphere of pure oxygen until all the aluminum had reacted to form an oxide. The oxide was collected and found to weigh 2.74 g. What is the empirical formula of the aluminum oxide?

*Analysis:* To obtain the empirical formula, we need to know the masses of aluminum and oxygen in a sample of the oxide. We know the oxide sample contains 1.45 g of Al, which was the amount of aluminum used. We also know the total mass of the oxide, which contains both oxygen and the 1.45 g of Al. By difference we can obtain the mass of oxygen in the oxide.

$$\text{mass of Al} = 1.45 \text{ g Al}$$

$$\text{mass of O} = 2.74 \text{ g} - 1.45 \text{ g} = 1.29 \text{ g O}$$

The procedure will be to calculate the moles of Al and O and then find the mole ratio and thus the empirical formula. The tools we will need are the atomic masses of Al and O.

$$1 \text{ mol Al} = 26.98 \text{ g Al}$$

$$1 \text{ mol O} = 16.00 \text{ g O}$$

*Solution:* The calculations are straightforward.

$$\text{moles of Al} = 1.45 \text{ g Al} \times \frac{1 \text{ mol Al}}{26.98 \text{ g Al}} = 0.0537 \text{ mol Al}$$

$$\text{moles of O} = 1.29 \text{ g O} \times \frac{1 \text{ mol O}}{16.00 \text{ g O}} = 0.0806 \text{ mol O}$$

Using the procedure in the text, we'll write the moles of each element as subscripts in the formula.

$$\text{Al}_{0.0537}\text{O}_{0.0806}$$

Then, in hope of finding whole numbers, we divide each subscript by the smallest one.

$$\text{Al}_{\frac{0.0537}{0.0537}} \text{O}_{\frac{0.0806}{0.0537}} = \text{Al}_{1.00}\text{O}_{1.50}$$

By dividing by the smallest subscript, we are sure that none of the resulting subscripts will be less than 1.

Well, we have one whole number subscript (for Al), but the other is not. The data is given to three significant figures, so we can't just round off the 1.50 to a whole number[13]. To obtain a set of whole numbers,

---

[13]The original mass data was given to three significant figures, so the subscripts we obtain should also have three significant figures. To round a subscript to a whole number, it should differ from a whole number by only a small amount in the third significant digit.

we'll use the trial-and-error method described in Example 3.12 on page 111 in the textbook. In this method, we first try multiplying both subscripts by 2 to see if we obtain a set of whole numbers. If that fails, then we try multiplying by 3, and if that fails, then by 4, and so on until we obtain a set of whole numbers.

Multiplying the subscripts by 2 gives

$$Al_{1.00 \times 2}O_{1.50 \times 2} \rightarrow Al_{2.00}O_{3.00}$$

Success! We've obtained whole-number subscripts, so the empirical formula is $Al_2O_3$.

***Is the Answer Reasonable?*** The fact that we've obtained subscripts that are whole numbers suggests that we've probably performed the calculations correctly.

---

One of the common ways chemical analysis data are presented is in the form of the percentages by mass of the elements in a compound. We can use these data to calculate empirical formulas as illustrated below.

**Empirical Formulas from Percentage Composition Data**

***PROBLEM 19***

It was determined that a compound of sulfur and fluorine was composed of 29.6% S and 70.4% F, by mass. What is the empirical formula of the compound?

***Analysis:*** Recall that the critical data needed to calculate an empirical formula are the masses of the elements in a sample of the compound. But we're not given masses in this problem. Before we can perform the empirical formula calculations, therefore, we must use the information given to obtain the required mass data.

If we had a known mass of the compound, we could use the percentages to calculate the mass of each element in the sample. Because the empirical formula doesn't depend on the sample size, we can imagine having any size sample we want, so for convenience, let's imagine we had a 100 g sample. The mass of sulfur in it would be 29.6 % of 100 g, which is 29.6 g S[14]. Similarly, the mass of fluorine in the 100 g sample would be 70.4% of 100 g, or 70.4 g F.

Now we can proceed as before, converting grams of sulfur and fluorine into moles of each and then establishing the mole ratio. To make these conversions, we use the atomic masses of sulfur and fluorine which allow us to set up the relationships

$$1 \text{ mol S} = 32.07 \text{ g S}$$

$$1 \text{ mol F} = 19.00 \text{ g F}$$

---

[14]Percent means "per hundred," so 29.6% S by mass means "29.6 g S per hundred grams compound." To take 29.6% of 100 g, we perform the following arithmetic,

$$100 \text{ g compound} \times \frac{29.6 \text{ g S}}{100 \text{ g compound}} = 29.6 \text{ g S}$$

*Solution:* First we convert grams of sulfur and fluorine to moles.

$$29.6 \text{ g S} \times \frac{1 \text{ mol S}}{32.07 \text{ g S}} = 0.923 \text{ mol S}$$

$$70.4 \text{ g F} \times \frac{1 \text{ mol F}}{19.00 \text{ g F}} = 3.71 \text{ mol F}$$

Now, we use these amounts as subscripts and then divide by the smallest.

$$S_{\frac{0.923}{0.923}}F_{\frac{3.71}{0.923}} = S_{1.00}F_{4.02}$$

The subscript we've calculated for fluorine is not exactly a whole number, but it differs from a whole number in the third and last significant digit. Because this digit could be a little larger or *smaller* than 2, we are justified in rounding 4.02 to 4.00, a whole number. The empirical formula, therefore, is $SF_4$.

*Is the Answer Reasonable?* As before, we've obtained whole-number subscripts, so the problem was probably solved correctly.

---

**Empirical Formulas from Indirect Analyses**

In the textbook we noted that it is usually not possible to decompose a compound into its elements in a way that allows us to weigh each element separately. Often, a chemical analysis is accomplished by causing the compound of interest to undergo chemical reactions that separate the elements in the compound from one another and capture them, without loss, in other compounds of known composition. These reaction products are weighed and the amounts of the elements in the original compound can then be computed. This is illustrated in the following example.

*PROBLEM 20*

A 1.071 g sample of a compound of carbon and sulfur was burned in oxygen. One of the products was sulfur dioxide, $SO_2$. The $SO_2$ was collected and found to weigh 1.80 g. Assuming that all the sulfur in the original sample was trapped in the $SO_2$, what is the empirical formula of the carbon-sulfur compound?

*Analysis:* To calculate the empirical formula, we need the masses of carbon and sulfur in the sample. This information isn't given in the problem, so we have to figure out how to get it.

We are assuming that all the sulfur in the original compound is contained in the $SO_2$ that was collected. Therefore, if we calculate the mass of sulfur in the 1.80 g $SO_2$, we will have the mass of sulfur that was in the original compound. We can then subtract this value from the total mass of 1.07 g to obtain the mass of carbon. Once we have the masses of sulfur and carbon, we can proceed as usual.

First, we need the tool to find the mass of sulfur in a sample of sulfur dioxide. We can obtain this from the data used to calculate the formula mass of $SO_2$.

| Element | Moles | Mass |
|---------|-------|------|
| sulfur | 1 mol | $1 \times 32.07 \text{ g} = 32.07 \text{ g S}$ |
| oxygen | 2 mol | $2 \times 16.00 \text{ g} = 32.00 \text{ g O}$ |
| | | Total        64.07 g $SO_2$ |

We can write the following.

$$64.07 \text{ g SO}_2 \Leftrightarrow 32.07 \text{ g S}$$

We will also need to use the atomic masses of sulfur and carbon to find the moles of each in the original compound.

$$1 \text{ mol S} = 32.07 \text{ g S}$$

$$1 \text{ mol C} = 12.01 \text{ g C}$$

Now that we have a plan and all the necessary tools, we can proceed with the solution.

**Solution:** First, we'll calculate the mass of sulfur in the $SO_2$.

$$1.80 \text{ g SO}_2 \times \frac{32.07 \text{ g S}}{64.07 \text{ g SO}_2} = 0.901 \text{ g S}$$

Now we can subtract this from the original mass of compound to find the mass of carbon in the sample.

$$\text{mass of C} = (1.071 \text{ g sample}) - (0.901 \text{ g S}) = 0.170 \text{ g C}$$

The calculations tell us that the compound in the original sample is composed of 0.901 g S and 0.170 g C. To find the empirical formula, we begin by converting these masses to moles.

$$0.170 \text{ g C} \times \frac{1 \text{ mol C}}{12.01 \text{ g C}} = 0.0142 \text{ mol C}$$

$$0.901 \text{ g S} \times \frac{1 \text{ mol S}}{32.07 \text{ g S}} = 0.0281 \text{ mol S}$$

Next, we write these moles as subscripts, and then divide by the smallest one.

$$C_{\frac{0.0142}{0.0142}}S_{\frac{0.0281}{0.0142}} = C_{1.00}S_{1.98}$$

The data for the masses of C and S have three significant figures, and the subscript of S differs slightly from a whole number in the third digit, so we are safe to round 1.98 to 2.00. The formula for the carbon-sulfur compound is $CS_2$.

**Is the Answer Reasonable?** Once again, the fact that we've obtained whole number subscripts after having done all these calculations gives us confidence we've solved the problem correctly.

The preceding problem is somewhat similar to Example 3.14 on page 113 of the textbook. Before we can do the work required to calculate the empirical formula, we must first find the masses of each element in the sample of the compound. In doing this preliminary work, it's important to understand that when we calculate these masses, they must be for the *same* sample or samples of the *same size*. Let's look at one more problem dealing with empirical formulas.

**PROBLEM 21**
A compound called cyanogen contains only the elements carbon and nitrogen. A 0.257 g sample of the compound was burned and the carbon in the sample was converted entirely into $CO_2$. When collected, the $CO_2$

weighed 0.435 g. A separate 0.353 g sample of cyanogen was treated in a way that converted all the nitrogen into ammonia, $NH_3$. The ammonia was collected and found to weigh 0.231 g. What is the empirical formula of cyanogen?

*Analysis:* We need to find the masses of carbon and nitrogen, but they must be for the same size cyanogen sample. The analyses used two samples of cyanogen, so if we find the mass of C in the $CO_2$ and the mass of N in the $NH_3$, they will be for samples of different sizes. How can we get around this dilemma?

One way is to use the data for one of the elements, let's say the nitrogen, to calculate the mass of $NH_3$ that would have been formed if the sample used to form it had been the same size as the one used to form the $CO_2$. In other words, we would like to solve the problem

$$0.257 \text{ g sample} \Leftrightarrow ? \text{ g } NH_3$$

From the data given for the second sample, we can write the following relationship.

$$0.353 \text{ g sample} \Leftrightarrow 0.231 \text{ g } NH_3$$

We can use it to find how much ammonia would have been formed if the sample size used in the analysis were 0.257 g. Let's do the calculation.

$$0.257 \text{ g sample} \times \frac{0.231 \text{ g } NH_3}{0.353 \text{ g sample}} = 0.168 \text{ g } NH_3$$

Now we know that a sample weighing 0.257 g will give 0.435 g $CO_2$ and 0.168 g $NH_3$. The procedure from this point will be the same as in the preceding problems. We first calculate the mass of C in the $CO_2$ and the mass of N in the $NH_3$. This will give us the masses of C and N in the same sized sample. The tools we will need are the masses of C in a mole of $CO_2$ and the mass of N in a mole of $NH_3$. Following the method shown before, the relationships are

$$12.01 \text{ g C} \Leftrightarrow 44.01 \text{ g } CO_2$$

$$14.01 \text{ g N} \Leftrightarrow 17.03 \text{ g } NH_3$$

After we calculate the masses of C and N, we convert them to moles using the atomic masses as tools.

$$1 \text{ mol C} = 12.01 \text{ g C}$$

$$1 \text{ mol N} = 14.01 \text{ g N}$$

Finally, we use the moles as subscripts and work to get whole-number subscripts.

*Solution:* We calculate the masses of C and N in the 0.257 g sample from the masses of $CO_2$ and $NH_3$.

$$0.435 \text{ g } CO_2 \times \frac{12.01 \text{ g C}}{44.01 \text{ g } CO_2} = 0.119 \text{ g C}$$

$$0.168 \text{ g } NH_3 \times \frac{14.01 \text{ g N}}{17.03 \text{ g } NH_3} = 0.138 \text{ g N}$$

Next, we convert these masses to moles.

$$0.119 \text{ g C} \times \frac{1 \text{ mol C}}{12.01 \text{ g C}} = 0.00991 \text{ mol C}$$

$$0.138 \text{ g N} \times \frac{1 \text{ mol N}}{14.01 \text{ g N}} = 0.00985 \text{ mol N}$$

Finally, we used these values as subscripts and divide by the smallest one.

$$C_{\frac{0.00991}{0.00985}} N_{\frac{0.00985}{0.00985}} = C_{1.01} N_{1.00}$$

Because the subscript for C differs from a whole number by a small amount in the last significant digit, we can safely round the subscript to 1.00. Therefore, the empirical formula of cyanogen is CN.

*Is the Answer Reasonable?*  After all the calculations we've done, it is unlikely that we would have obtained whole number subscripts unless we had solved the problem correctly. Therefore, we should feel confident in our result.

---

**Finding Molecular Formulas**

The subscripts in a molecular formula are either equal to those in the empirical formula, or they are a multiple of them. For example, consider butane. Its empirical formula is $C_2H_5$ and its molecular formula is $C_4H_{10}$. Each subscript in the molecular formula is twice as large as the corresponding subscript in the empirical formula. A consequence of this is that the molecular mass of $C_4H_{10}$ (58.12 g mol$^{-1}$) is twice the formula mass of $C_2H_5$ (29.06 g mol$^{-1}$). This observation provides the basis for finding molecular formulas from empirical formulas, *the critical link being the measured molecular mass of the compound*[15].

In general, dividing the molecular mass by the empirical formula mass yields the factor by which the subscripts in the empirical formula are multiplied to obtain those in the molecular formula. Once again, consider butane. Suppose we only know the empirical formula, $C_2H_5$ and the molecular mass (58.12 g mol$^{-1}$). To find the molecular formula, we first divide the molecular mass of the compound (58.12 g mol$^{-1}$) by the empirical formula mass (29.06 g mol$^{-1}$).

$$\frac{58.12 \text{ g mol}^{-1}}{29.06 \text{ g mol}^{-1}} = 2$$

We then multiply each of the subscripts in the empirical formula by the result (in this case, 2).

$$C_2H_5$$
$$\downarrow \quad \downarrow$$
$$\times 2 \quad \times 2$$
$$\downarrow \quad \downarrow$$
$$C_4H_{10}$$

---

[15]Experimental methods for measuring molecular masses are discussed later in the course. For now, we will provide the molecular mass as part of the data in problems that ask you to determine a molecular formula from an empirical formula.

## PROBLEM 22

The molecular mass of cyanogen is measured to be 52.1 g mol$^{-1}$. The empirical formula of the compound is CN. What is the molecular formula?

*Analysis:* This problem doesn't take a lot of deep thinking. The procedure is straight forward. To obtain the molecular formula we need two pieces of data, the empirical formula and the molecular mass. We have both of these, so enough information is available to solve the problem.

*Solution:* For CN, the empirical formula mass (i.e., the formula mass corresponding the empirical formula) is easily calculated to be 12.01 + 14.01 = 26.02. We divide the molecular mass by this value.

$$\frac{52.1}{26.02} = 2.00$$

To obtain the molecular formula, we multiply each of the subscripts in the empirical formula by 2.

$$C_{1\times2}N_{1\times2} = C_2N_2$$

The molecular formula of cyanogen is $C_2N_2$.

*Is the Answer Reasonable?* There isn't much we can do here except note that the ratio of the molecular mass to the empirical formula mass is a whole number, which it is expected to be.

---

**Additional Problems**

1. 2.04 g of carbon reacts with 5.44 g of oxygen to form 7.48 g of compound. How many atoms of O per atom of C are in this compound?

2. What is the empirical formula of a purified drug compound that is 74.27% C, 7.79% H, 12.99% N, and 4.95% O?

3. Elemental analysis of a pure compound isolated from tea leaves gave the following results: 49.48% C, 5.19% H, 28.85% N, 16.48% O. What is the empirical formula of the compound?

4. 1.000 g of an unknown compound containing carbon, hydrogen, and oxygen is completely burned in pure oxygen. 1.47 g of $CO_2$ and 0.60 g of $H_2O$ are collected. If all of the carbon in the original sample was trapped in the $CO_2$, and all of the hydrogen appears as $H_2O$ after the reaction, what was the empirical formula of the unknown compound?

5. A compound has an empirical formula of CH and a molecular mass of 78.11 g mol$^{-1}$. What is its molecular formula?

**Answers**

1. $CO_2$, 2. $C_{20}H_{25}N_3O$, 3. $C_4H_5N_2O$, 4. $CH_2O$, 5. $C_6H_6$

---

### Balanced Chemical Equations

**Skills Analysis**

1. Balance the following equations. Make all of the coefficients in the balanced equation integers.

(a) __ $CH_4$ + __ $O_2 \rightarrow$ __ $CO_2$ + __ $H_2O$

(b) __ $HCl$ + __ $Ba(OH)_2 \rightarrow$ __ $H_2O$ + __ $BaCl_2$

(c) __ $K_2SO_4$ + __ $AgNO_3$ → __ $Ag_2SO_4$ + __ $KNO_3$

(d) __ Eu + __ $O_2$ → __ $Eu_2O_3$

(e) __ $C_6H_{12}O_6$ + __ $O_2$ → __ $CO_2$ + __ $H_2O$

**2.** How many moles of carbon dioxide are released when 25 moles of octane ($C_8H_{18}$) are burned? The combustion of octane can be written

$$2\,C_8H_{18} + 25\,O_2 → 16\,CO_2 + 18\,H_2O$$

**3.** Baking soda ($NaHCO_3$) smothers fires by creating carbon dioxide gas:

$$2\,NaHCO_3 → Na_2CO_3 + H_2O + CO_2$$

How many moles of $CO_2$ will be produced from 1.00 kg of $NaHCO_3$?

**4.** Baking soda causes baked goods to rise by a different reaction:

$$NaHCO_3 + H^+ → Na^+ + H_2O + CO_2$$

How many grams of $CO_2$ will be produced from 10.0 grams of $NaHCO_3$ in this reaction?

**5.** Iodine ($I_2$) can be prepared from sodium iodide (NaI) by the following reaction:

$$Cl_2 + 2\,NaI → I_2 + 2\,NaCl$$

How many grams of NaI would be needed to completely react with 1.00 g of $Cl_2$?

If you missed question 1, study the material on balancing equations in Chapter 3 of the text before you attempt this section. If you missed question 2, you should read this entire section. If you missed questions 3, 4, or 5, carefully study the last three examples given in this section.

**1.** (a) $1CH_4 + 2O_2 → 1CO_2 + 2H_2O$, (b) $2HCl + 1Ba(OH)_2 → 2H_2O + 1BaCl_2$, (c) $1K_2SO_4 + 2AgNO_3 → 1Ag_2SO_4 + 2KNO_3$, (d) $4Eu + 3O_2 → 2Eu_2O_3$, (e) $1C_6H_{12}O_6 + 3O_2 → 6CO_2 + 6H_2O$, **2.** 200 mol $CO_2$, **3.** 5.95 mol $CO_2$, **4.** 5.24 g $CO_2$, **5.** 4.23 g NaI

For stoichiometry problems dealing with chemical reactions, balanced chemical equations are essential[16]. This is because the coefficients in an equation provide the *only way* to relate the relative amounts of reactants and products. This relationship is on both a molecular and mole basis. Consider, for example, the balanced equation for the combustion of methanol ($CH_3OH$), the fuel in "canned heat" products such as Sterno.

**Balanced Chemical Equations as Mole-to-Mole Critical Links for Chemical Reactions**

$$2CH_3OH + 3O_2 → 2CO_2 + 4H_2O$$

On a molecular basis, we see that 2 molecules of $CH_3OH$ combine with 3 molecules of $O_2$ to form 2 molecules of $CO_2$ and 4 molecules of $H_2O$. Scaling up to laboratory amounts, we find that when 2 moles of $CH_3OH$ burn, they consume 3 moles of $O_2$ and form 2 moles of $CO_2$ and 4 moles

---

[16]In Chapter 3 you are taught to balance equations by inspection. It is a skill you should be sure to practice by working Review Problems 3.87–3.90 on page 144 of the textbook.

of $H_2O$. Notice that the *mole ratios* among reactants and products are numerically equal to the ratios expressed in terms of the individual molecules of the reactants and products.

When we face a stoichiometry problem involving substances in a chemical reaction, the critical links are the coefficients of the balanced equation. These coefficients *alone* provide the relationships among moles of reactants and moles of products. For example, if we wished to relate the amounts of $CO_2$ and $H_2O$ formed in the combustion of $CH_3OH$, we would use the coefficients of $CO_2$ and $H_2O$ to establish the mole relationship

$$2CH_3OH + 3O_2 \rightarrow 2CO_2 + 4H_2O$$

$$2 \text{ mol } CO_2 \Leftrightarrow 4 \text{ mol } H_2O$$

Similarly, if we wished to relate the amount of $O_2$ consumed when a certain amount of $CO_2$ is formed in the reaction, we would use the coefficients of $O_2$ and $CO_2$ in the equation to write

$$3 \text{ mol } O_2 \Leftrightarrow 2 \text{ mol } CO_2$$

Let's look at some sample problems of varying complexity.

### PROBLEM 23

In a gas barbecue, the combustion of propane, $C_3H_8$, follows the equation

$$C_3H_8(g) + 5O_2(g) \rightarrow 3CO_2(g) + 4H_2O(g)$$

How many moles of water vapor are formed if 0.250 mol of $C_3H_8$ is burned?

*Analysis:* Let's begin by stating the problem in equation form.

$$0.250 \text{ mol } C_3H_8 \Leftrightarrow ? \text{ mol } H_2O$$

To solve the problem we need to find a relationship between the amounts of these substances. Because the $C_3H_8$ and $H_2O$ are involved in the combustion reaction, the tools we need are their coefficients in the balanced equation. The coefficients of $C_3H_8$ and $H_2O$ give us the following

$$1 \text{ mol } C_3H_8 \Leftrightarrow 4 \text{ mol } H_2O$$

This provides the connection we need to convert moles of $C_3H_8$ to moles of $H_2O$.

*Solution:* To solve the problem we apply the mole relationship between $C_3H_8$ and $H_2O$ as a conversion factor.

$$0.250 \text{ mol } C_3H_8 \times \frac{4 \text{ mol } H_2O}{1 \text{ mol } C_3H_8} = 1.00 \text{ mol } H_2O$$

We conclude that combustion of 0.250 mol $C_3H_8$ will produce 1.00 mol $H_2O$.

***Is the Answer Reasonable?*** This is a pretty simple problem, and you could probably see the answer without having to work it out in such great detail. The coefficients tell us that the number of moles of water formed is four times the number of moles of propane consumed. Therefore, when 0.250 mol $C_3H_8$ is burned, we will get $4 \times 0.250$ mol = 1.00 mol of water as one of the products.

## PROBLEM 24

Calcium carbonate, $CaCO_3$ (the principal ingredient in the antacid Tums), reacts with hydrochloric acid, HCl (stomach acid), according to the following equation.

$$CaCO_3(s) + 2HCl(aq) \rightarrow CaCl_2(aq) + CO_2(g) + H_2O$$

How many moles of HCl react with 0.250 g of $CaCO_3$?

***Analysis:*** We'll begin by expressing the problem in the form of an equation

$$0.250 \text{ g } CaCO_3 \Leftrightarrow \text{ ? mol HCl}$$

We are dealing with amounts of substances involved in a chemical reaction, so we will have to use their coefficients to relate moles. Using the coefficients of $CaCO_3$ and HCl, we can write

$$1 \text{ mol } CaCO_3 \Leftrightarrow 2 \text{ mol HCl}$$

We have the tool needed to go from moles of $CaCO_3$ to the units of the answer, moles of HCl. We now need a tool to convert grams of $CaCO_3$ to moles of $CaCO_3$. The tool, of course, is the formula mass of $CaCO_3$. Adding up the atomic masses gives a formula mass of 100.1, so we can write

$$1 \text{ mol } CaCO_3 = 100.1 \text{ g } CaCO_3$$

Now we have all the information we need to go from grams of $CaCO_3$ to moles of HCl, so we can proceed with the solution.

***Solution:*** We use the relationships we've established,

$$1 \text{ mol } CaCO_3 \Leftrightarrow 2 \text{ mol HCl}$$

$$1 \text{ mol } CaCO_3 = 100.1 \text{ g } CaCO_3$$

to form conversion factors (add the cancel marks yourself).

$$0.250 \text{ g } CaCO_3 \times \frac{1 \text{ mol } CaCO_3}{100.1 \text{ g } CaCO_3} \times \frac{2 \text{ mol HCl}}{1 \text{ mol } CaCO_3} = 0.00500 \text{ mol HCl}$$

The amount of HCl consumed will be 0.00500 mol.

***Is the Answer Reasonable?*** Let's do some approximate arithmetic. The number of moles of $CaCO_3$ equals the mass of $CaCO_3$ divided by its formula mass (0.25 g divided by approximately 100 g $mol^{-1}$). The result is 0.0025 mol of $CaCO_3$. Two moles of HCl are consumed for every one mole of $CaCO_3$ that reacts, so the amount of HCl consumed will be $2 \times 0.0025 = 0.005$ mol. This agrees with our answer, so the answer appears to be correct.

**PROBLEM 25**

Magnesium hydroxide, $Mg(OH)_2$ (the creamy white substance in milk of magnesia), reacts with hydrochloric acid, HCl (found in stomach acid), to give magnesium chloride and water. The reaction is

$$Mg(OH)_2(s) + 2HCl(aq) \rightarrow MgCl_2(aq) + 2H_2O$$

How many grams of $Mg(OH)_2$ are required to react completely with 12.8 g of HCl?

*Analysis:* We'll begin by expressing the problem in the form of an equation.

$$12.8 \text{ g HCl} \Leftrightarrow ? \text{ g Mg(OH)}_2$$

We see that we are attempting to relate amounts of substances involved in a chemical reaction. This immediately suggests that we will need to use their coefficients in the balanced equation to relate them on a mole basis. Let's write that relationship.

$$1 \text{ mol Mg(OH)}_2 \Leftrightarrow m\ 2 \text{ mol HCl}$$

We have part of the solution, but we are still missing some connections.

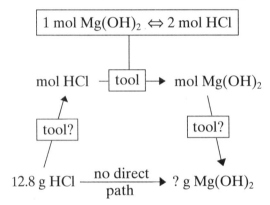

There is not direct path between grams of HCl and grams of $Mg(OH)_2$, but we have the path between moles. To complete the connections between units, we need to relate grams to moles for both substances. By now you should realize that the tool to accomplish this is the formula mass. For HCl, the formula mass is 36.46 and for $Mg(OH)_2$ it is 58.32. Therefore, we can write

$$1 \text{ mol HCl} = 36.46 \text{ g HCl}$$

$$1 \text{ mol Mg(OH)}_2 = 58.32 \text{ g Mg(OH)}_2$$

Now we have a complete path from the units given (g HCl) to the units desired (g $Mg(OH)_2$), so we can proceed to the solution step.

*Solution:* We have all the relationships we need, so we apply them as conversion factors, being sure the units cancel.

$$12.8 \text{ g HCl} \times \frac{1 \text{ mol HCl}}{36.46 \text{ g HCl}} \times \frac{1 \text{ mol Mg(OH)}_2}{2 \text{ mol HCl}} \times \frac{58.32 \text{ g Mg(OH)}_2}{1 \text{ mol Mg(OH)}_2}$$

$$= 10.2 \text{ g Mg(OH)}_2$$

The calculations tell us that 12.8 g of HCl will require 10.2 g of $Mg(OH)_2$ for complete reaction.

***Is the Answer Reasonable?*** We can do some proportional reasoning and approximate arithmetic to get an idea whether our answer is "in the right ballpark." The equation tells us that 2 mol HCl are needed to react with 1 mol $Mg(OH)_2$. The mass of 2 mol HCl is approximately 2 $\times$ 36 = 72 g, and the mass of 1 mol of $Mg(OH)_2$ is approximately 60 g. So we expect that 72 g of HCl will react with approximately 60 g of $Mg(OH)_2$. In other words, the mass of $Mg(OH)_2$ consumed will be a little less than the mass of HCl that reacts. Our answer, 10.2 g $Mg(OH)_2$, is a little less than 12.8 g HCl, so the answer seems to be reasonable.

The preceding problem demonstrates an important lesson about problems that deal with chemical reactions. The critical connections among the substances involved in a reaction are mole relationships provided by the coefficients of the balanced equation. Notice that this was the first relationship we established. Then we proceeded to find connections that take us from: (1) the given data to moles of the first substance, and (2) from moles of the second substance to the units of the desired answer. The following "flow diagram" shows the tools we used in the various steps in solving the problem.

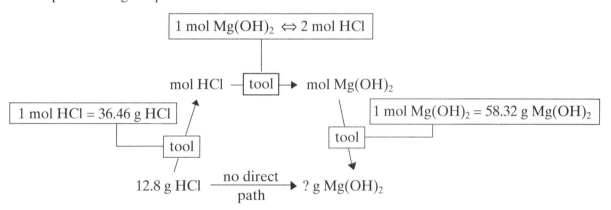

With practice, you will find that all stoichiometry problems involving chemical reactions are pretty much the same. The first step is to establish the balanced chemical equation. Then, using the coefficients, write the mole relationship between the substances of interest. After this, look for any additional connections you may need between moles and the units of the given data and/or moles and the units of the answer. Once you've reached this point, the hard work is done. All that's left is to assemble the relationships into conversion factors and calculate the answer.

**1.** Telegraph stations in the 1800s generated electricity with batteries that use the following reaction:   **Additional Problems**

$$Zn + CuSO_4 \rightarrow Cu + ZnSO_4$$

For each 1.00 gram of Zn consumed, how many grams of copper appear?

**2.** Potassium (K) reacts violently with water to produce hydrogen gas ($H_2$)

$$2K + 2H_2O \rightarrow 2KOH + H_2$$

How many moles of hydrogen gas will be produced when a pellet of potassium weighing 0.100 g is dropped into a bucket of water?

**3.** A spot test for sulfate ions is based on the following reaction:

$$BaCl_2 + SO_4^{2-} \rightarrow BaSO_4(s) + 2Cl^-$$

How many grams of $BaCl_2$ are needed to produce 1.00 mg of $BaSO_4$?

**4.** Silver nitrate ($AgNO_3$) can be used to analyze samples for sodium chloride content by the following reaction:

$$AgNO_3 + NaCl \rightarrow AgCl(s) + NaNO_3$$

If a sample treated with excess silver nitrate produced 1.000 g of AgCl, how many grams of NaCl were in the sample?

**5.** A 1.0000 g sample of poison containing NaF is analyzed using the following reaction:

$$2NaF + Ca(NO_3)_2 \rightarrow CaF_2(s) + 2NaNO_3$$

If treatment of the sample with excess $Ca(NO_3)_2$ produced 0.0930 g of $CaF_2$ crystals, what is the percentage of NaF in the poison?

**Answers**

**1.** 0.972 g Cu, **2.** 0.00128 mol $H_2$, **3.** 0.000892 g $BaCl_2$, **4.** 0.4078 g NaCl, **5.** 10.0% NaF

## Limiting Reactant Calculations

**Skills Analysis**

**1.** Hydrogen chloride gas reacts with ammonia gas to produce ammonium chloride:

$$HCl + NH_3 \rightarrow NH_4Cl$$

If 10 molecules of HCl and 20 molecules of $NH_3$ are placed in a container, how many molecules of HCl and $NH_3$, and formula units of $NH_4Cl$ will be in the container when the reaction is complete?

**2.** Ammonia is synthesized using the following reaction:

$$N_2 + 3H_2 \rightarrow 2NH_3$$

If 5.0 moles of $N_2$ are added to 10.0 moles of $H_2$, how many moles of $NH_3$ can be produced?

**3.** Cobalt violet, $Co_3(PO_4)_2$, is an artist's pigment which can be prepared using the following reaction:

$$3CoCl_2 + 2Na_3PO_4 \rightarrow Co_3(PO_4)_2 + 6NaCl$$

(a) How many grams of cobalt violet can be prepared from 1.00 g of $CoCl_2$ and 1.00 g of $Na_3PO_4$, if they react completely?

(b) How much excess reactant remains after the reaction is complete?

If you miss any of the questions, read the entire section.
**1.** The container will hold 10 molecules of $NH_3$ and 10 formula units of $NH_4Cl$ when the reaction is complete. HCl is the limiting reactant, and is completely consumed. **2.** 6.67 mol $NH_3$, **3.** (a) 0.941 g $Co_3(PO_4)_2$, (b) 0.16 g $Na_3PO_4$ left over.

When substances are combined for a chemical reaction, they are not always mixed together in just the right proportions to get complete reaction. As a result, sometimes one of the reactants will be completely used up before the rest, and once this happens, no further products can form. The reaction will simply come to a halt. At this point, the reaction mixture will contain the products of the reaction plus some of the reactant that had not been able to react. The following diagram illustrates this for the reaction of carbon with oxygen to form carbon dioxide.

$$C + O_2 \rightarrow CO_2$$

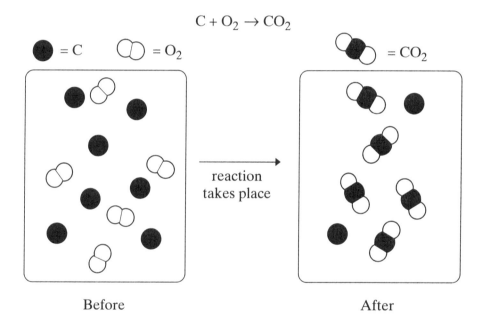

Before                                    After

Notice that in both the before and after views, there are the same numbers of atoms of C and O. However, there are not enough oxygen atoms to combine with all the carbon, so after the oxygen is all used up, no more $CO_2$ can form. At the end of the reaction, there is still some unreacted carbon along with the $CO_2$ that has formed.

In this reaction mixture, we call the oxygen the **limiting reactant** because we don't have enough of it to consume all the carbon; its amount is what *limits* the amount of product ($CO_2$) that forms. We might call the carbon the **excess reactant**, because it is present in *excess* amount—that is, in an amount greater than is needed to react with all the oxygen.

If we wish to predict the amount of product in a reaction, we need to know which of the reactants is the limiting reactant. For example, consider the following mixture of C and $O_2$. How many molecules of $CO_2$ can form from this mixture?

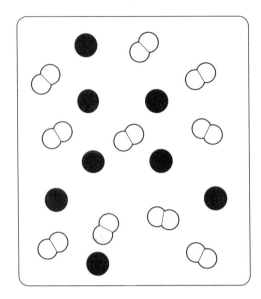

In this mixture, we have 10 molecules of $O_2$ and 8 atoms of C. It's pretty easy to see which is the limiting reactant. We can figure it out two ways (although we only need to do it one way if all we need to know is which is limiting). The reasoning we do here will be used again later, so be sure you understand it.

One way to find the limiting reactant is to examine the $O_2$. Each $O_2$ requires one C to form $CO_2$, so 10 $O_2$ molecules would require 10 carbon atoms. We have only 8 C atoms, so C must be the limiting reactant. It will be the one used up first and its amount limits the amount of $CO_2$ that can form.

The second way is to look at the carbon. Each carbon requires one $O_2$ molecule. Therefore, 8 C atoms require 8 $O_2$ molecules. We have more than 8 $O_2$ molecules, so there is an excess of $O_2$. If $O_2$ is present in excess, then carbon must be the limiting reactant.

Once we've identified the limiting reactant, we can figure out how much product will form and how much of which reactant is left over. *We always use the limiting reactant to determine how much product is formed.* When all 8 atoms of C react, they will yield 8 molecules of $CO_2$, so this is the amount of product that forms. There will be $O_2$ left over, and the amount left over is the difference between the amount of $O_2$ originally available (10 molecules) and the amount that reacts with the carbon (8 molecules). This difference, 2 molecules, is the amount of $O_2$ left over.

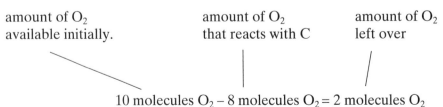

amount of $O_2$ available initially.

amount of $O_2$ that reacts with C

amount of $O_2$ left over

10 molecules $O_2$ – 8 molecules $O_2$ = 2 molecules $O_2$

There is a particular strategy we will employ in solving limiting reactant problems, but before we can use it, we must be able to recognize when a problem falls into this class. The key is in the way the problem is worded. The following are two stoichiometry problems that deal with the same chemical reaction, the reaction of $Mg(OH)_2$ with HCl.

$$Mg(OH)_2 + 2HCl \rightarrow MgCl_2 + 2H_2O$$

### PROBLEM 26
How many grams of $MgCl_2$ can be formed when 2.48 g of $Mg(OH)_2$ reacts with HCl?

### PROBLEM 27
How many grams of $MgCl_2$ can be formed when 2.48 g of $Mg(OH)_2$ are combined with 2.88 g of HCl?

In Problem 26, we are not told about the amount of HCl available, so the only way we can solve the problem is to assume that there is more than enough HCl available. We would therefore base the calculation on the amount of $Mg(OH)_2$ given in the problem. Problem 26 is *not* a limiting reactant problem.

Problem 27 *is* a limiting reactant problem. *Notice that we are given amounts of both reactants*; this is the clue that identifies it as a limiting reactant problem. We can't tell, reading the problem, which one of the reactants will be completely used up, and therefore we don't know at this point which reactant to use to calculate the amount of product. To solve the problem, we must first identify the limiting reactant and then use the amount of that reactant to calculate the amount of product that can form.

The approach we will take in solving limiting reactant problems is as follows:

**Strategy for Solving Limiting Reactant Problems**

1. We will first determine the number of moles of both reactants.

2. We will select one reactant and calculate the amount of the other that is required for complete reaction. This will require the coefficients of the balanced equation.

3. We will compare the amount needed for complete reaction with the amount available and determine which of the reactants is limiting. We described this reasoning above in our example of the reaction of C with $O_2$ to form $CO_2$. Review it if necessary.

4. We will use the amount of limiting reactant to calculate the amount of product formed. We can also calculate the amount of the excess reactant that is actually consumed and, by difference, determine the amount of excess reactant left over.

With this as background, let's work on Problem 27, which we'll restate.

## PROBLEM 27

How many grams of $MgCl_2$ can be formed when 2.48 g of $Mg(OH)_2$ are combined with 2.88 g of HCl?

$$Mg(OH)_2 + 2HCl \rightarrow MgCl_2 + 2H_2O$$

***Analysis and Solution:*** You already know this is a limiting reactant problem, but in other situations, you should look for the clue that identifies it as a problem of this type—namely, that we are given amounts of both reactants. Then we proceed as described in the strategy. We'll combine the Analysis and Solution steps here because we can't simply assemble all the data and perform the calculations all at once.

The first step is to determine the number of moles of the reactants that are available. We are given grams, so we convert to moles using the formula masses. The relationships are

$$1 \text{ mol HCl} = 36.46 \text{ g HCl}$$

$$1 \text{ mol } Mg(OH)_2 = 58.32 \text{ g } Mg(OH)_2$$

Applying them, gives

$$2.48 \text{ g } Mg(OH)_2 \times \frac{1 \text{ mol } Mg(OH)_2}{58.32 \text{ g } Mg(OH)_2} = 0.0425 \text{ mol } Mg(OH)_2$$

$$2.88 \text{ g HCl} \times \frac{1 \text{ mol HCl}}{36.46 \text{ g HCl}} = 0.0790 \text{ mol HCl}$$

Next, we select one reactant and calculate the number of moles of the other required for complete reaction. For this step we need the coefficients of the balanced equation, from which we obtain

$$1 \text{ mol } Mg(OH)_2 \Leftrightarrow 2 \text{ mol HCl}$$

It doesn't matter which reactant we choose, so let's select the HCl and calculate how many moles of $Mg(OH)_2$ are needed to use it up.

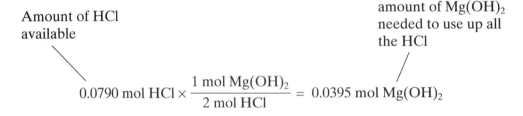

$$0.0790 \text{ mol HCl} \times \frac{1 \text{ mol } Mg(OH)_2}{2 \text{ mol HCl}} = 0.0395 \text{ mol } Mg(OH)_2$$

Now we compare the amount of $Mg(OH)_2$ that's available in the reaction mixture (0.0425 mol) with the amount of $Mg(OH)_2$ that's required to react with all the HCl.

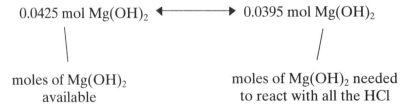

Notice that there is more $Mg(OH)_2$ available than is needed to use up all the HCl. This means that $Mg(OH)_2$ is the reactant in excess and that some $Mg(OH)_2$ will be left over. It also means that *HCl must be the limiting reactant.*

Now that we've identified the limiting reactant, we can calculate the amount of $MgCl_2$ that can form. We use the moles of HCl to calculate the moles of $MgCl_2$ that can form, and then convert that to grams.

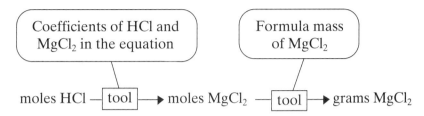

The chemical equation gives us

$$2 \text{ mol HCl} \Leftrightarrow 1 \text{ mol MgCl}_2$$

and the formula mass of $MgCl_2$ gives us

$$1 \text{ mol MgCl}_2 = 95.2 \text{ g MgCl}_2$$

We use these and the amount of the limiting reactant to calculate the mass of $MgCl_2$ that could be formed in the reaction.

$$0.0790 \text{ mol HCl} \times \frac{1 \text{ mol MgCl}_2}{2 \text{ mol HCl}} \times \frac{95.2 \text{ g MgCl}_2}{1 \text{ mol MgCl}_2} = 3.76 \text{ g MgCl}_2$$

The results tell us that the maximum amount of $MgCl_2$ that could form in the reaction using these amounts of reactants is 3.76 g.

***Is the Answer Reasonable?*** There are a few things we can check without too much trouble. Let's start with the number of moles of $Mg(OH)_2$ and HCl in the reaction mixture. For $Mg(OH)_2$, 0.1 mol would weigh 5.8 g, so 2.5 g is about half of 0.1 mol, or 0.05 mol; the value we obtained (0.0425 mol) seems about right. Similarly, 0.1 mol HCl weighs about 3.6 g, so 2.88 g HCl is a little less than 0.1 mol. The value we obtained (0.0790 mol) seems okay.

We can check the limiting reactant conclusion by working with the $Mg(OH)_2$. If 0.0425 mol $Mg(OH)_2$ were to react, it would required twice as many moles of HCl, or 0.0850 mol HCl. We have only 0.0790 mol HCl, so there isn't enough HCl to react with all the $Mg(OH)_2$. The reactant we "don't have enough of" is the limiting reactant, so once again, we conclude that HCl is the limiting reactant.

According to the coefficients in the equation, the number of moles of $MgCl_2$ that could form should equal half the number of moles of HCl that reacts. Half of 0.079 is about 0.04, so we expect about 0.04 moles of $MgCl_2$ could form. The formula mass of $MgCl_2$ is 95.2, so a mole of $MgCl_2$ weighs about 100 g. Therefore, 0.04 mol of $MgCl_2$ would weigh about 4 g. Our answer of 3.76 is not far from this, so the answer does seem to be reasonable.

In the preceding problem, we could also have been asked to calculate the amount of the excess reactant that would remain after the reaction is complete. This is a fairly simple calculation. We determined that the HCl would consume 0.0395 mol of $Mg(OH)_2$. We also know that we begin with 0.0425 mol $Mg(OH)_2$. The difference between these two values is the amount of $Mg(OH)_2$ that will be left over.

$$0.0425 \text{ mol } Mg(OH)_2 - 0.0395 \text{ mol } Mg(OH)_2 = 0.0030 \text{ mol } Mg(OH)_2$$

Amount available in the reaction mixture.

Amount that would react with the HCl.

Amount left over.

The amount of $Mg(OH)_2$ left over would be 0.0030 mol. If we wished to know its mass, we multiply by the mass of a mole of $Mg(OH)_2$.

$$0.0030 \text{ mol } Mg(OH)_2 \times \frac{58.32 \text{ g } Mg(OH)_2}{1 \text{ mol } Mg(OH)_2} = 0.17 \text{ g } Mg(OH)_2$$

**Additional Problems**

1. Oil of wintergreen (methyl salicylate, $HOC_6H_4COOCH_3$) is synthesized from salicylic acid ($HOC_6H_4COOH$) and methanol ($CH_3OH$) as follows:

$$HOC_6H_4COOH + CH_3OH \rightarrow HOC_6H_4COOCH_3 + H_2O$$

If 1.00 g of $HOC_6H_4COOH$ is added to 1.00 g $CH_3OH$, what is the maximum mass of $HOC_6H_4COOCH_3$ that can be obtained, in grams?

2. Ethanol ($C_2H_5OH$) is synthesized for industrial use by the following reaction, carried out at very high pressure:

$$C_2H_4(g) + H_2O(g) \rightarrow C_2H_5OH$$

How many grams of ethanol can be produced when 1.00 kg of ethylene ($C_2H_4$) and 1.00 kg of steam react completely?

3. Some of the acid in acid rain is produced by the following reaction:

$$3NO_2(g) + H_2O(l) \rightarrow 2HNO_3(aq) + NO(g)$$

Suppose that a falling raindrop weighing 0.0500 g absorbs 1.00 µg of $NO_2$. How many micrograms of $HNO_3$ can be produced?

**Answers**

1. 1.10 g $HOC_6H_4COOCH_3$, 2. 1640 g $C_2H_5OH$, 3. 0.913 µg $HNO_3$

**Relating Moles of Solute and Volume of Solution**

**Skills Analysis**

1. What is the molarity of NaCl in 4.0 L of a solution that contains 2.0 mol of NaCl?

2. How many milliliters of a 1.0 $M$ NaCl solution will contain 2.0 mmol of NaCl?

3. How many grams of NaCl are in 100.0 mL of a 2.00 $M$ NaCl solution?

4. Tea contains as much as 50 mg of caffeine in a 150 mL cup. What is the molarity of caffeine in tea? (Caffeine is $C_8H_{10}N_4O_2$, with a molecular mass of 194.19 g $mol^{-1}$.)

**5.** How many milliliters of coffee would deliver exactly 200 mg of caffeine, if the molarity of caffeine in coffee is $7.60 \times 10^{-3}$ $M$?

**6.** How many grams of caffeine would be needed to prepare 1.0 L of a solution that is 0.020 $M$ in caffeine?

If you missed any of these questions, you should read the entire section. If you answered them correctly by converting mmol to mol and liters to milliliters, you may want to study "Equivalent Ways of Expressing Molar Concentration" for a more efficient way to solve these problems.

**1.** 0.5 $M$ NaCl, **2.** 2.0 mL, **3.** 11.7 g NaCl, **4.** $1.7 \times 10^{-3}$ $M$, **5.** 136 mL coffee **6.** 3.9 g caffeine

It is quite common to carry out reactions in solution. By dissolving the reactants in a solvent before combining them, the reactant particles (atoms, molecules, or ions) are able to mix completely and the reaction is able to proceed smoothly and quickly.

The composition of a solution is variable, meaning the proportions of solute to solvent can vary. To express the composition of a solution, we specify the *concentration* of the solute, and for purposes of stoichiometry, one of the most convenient expressions of concentration is molarity. **Molarity** is defined as the ratio of the number of moles of solute to the volume of the solution expressed in liters.

$$\text{molarity } (M) = \frac{\text{moles of solute (mol)}}{\text{liters of solution (L)}}$$

Thus, a solution that contains 0.25 mole of solute per liter of solution has a concentration that we can express as 0.25 mol $L^{-1}$, or 0.25 $M$. The symbol $M$ stands for the units mol $L^{-1}$. For example, a solution that is labeled 0.10 $M$ HCl has a concentration of HCl equal to 0.10 mol $L^{-1}$.

To calculate the molarity of a solution, we need only the number of moles of solute (or the mass of solute, which we can convert to moles) and the volume of the solution. For instance, suppose we had 275 mL of a solution in which is dissolved 0.0264 mol of $CaCl_2$. To calculate the molarity, we divide the number of moles by the volume expressed in liters.

$$\text{molarity} = \frac{0.0264 \text{ mol } CaCl_2}{0.275 \text{ L solution}} = 0.0960 \ M \ CaCl_2$$

> Converting from milliliters to liters involves moving the decimal three places to the left. 275 mL = 0.275 L

Notice how we converted milliliters to liters. You should practice converting between these two units (from mL to L and from L to mL) until it becomes effortless. It is a skill you will use often.

**Equivalent Ways of Expressing Molar Concentration**

Because molarity is a ratio, it is possible to express the ratio in units other than moles and liters, and sometimes it's convenient to do so. One alternative is to convert the unit liter in the denominator to milliliters. Because 1 L = 1000 mL, we can express molarity in the units "mol/1000 mL". Thus, the concentration 0.20 $M$ HCl can be expressed as follows:

$$0.20\ M\ \text{HCl} = \frac{0.20\ \text{mol HCl}}{1\ \text{L solution}} = \frac{0.20\ \text{mol HCl}}{1000\ \text{mL solution}}$$

Another alternative is to convert the numerator to millimoles and the denominator to milliliters using: 1 mmol = $10^{-3}$ mol and 1 mL = $10^{-3}$ L. Let's do this for the 0.20 $M$ HCl solution.

$$0.20\ M = \frac{0.20\ \text{mol HCl} \times \left(\dfrac{1\ \text{mmol HCl}}{10^{-3}\ \text{mol HCl}}\right)}{1\ \text{L solution} \times \left(\dfrac{1\ \text{mL solution}}{10^{-3}\ \text{L solution}}\right)} = \frac{0.20 \times \dfrac{1}{10^{-3}}\ \text{mmol HCl}}{1 \times \dfrac{1}{10^{-3}}\ \text{mL solution}}$$

Notice that the result gives us the quantity $1/10^{-3}$ in both numerator and denominator. When we cancel this, we obtain 0.20 mmol HCl/1 mL solution

$$0.20\ M = \frac{0.20 \times \dfrac{1}{10^{-3}}\ \text{mmol HCl}}{1 \times \dfrac{1}{10^{-3}}\ \text{mL solution}} = \frac{0.20\ \text{mmol HCl}}{1\ \text{mL solution}}$$

Thus, the ratio of millimoles to milliliters is equal to the ratio of moles to liters. This means we can express molarity in either of these two sets of units. For example, if we had a 0.10 $M$ solution of NaOH, we could write

$$0.10\ M\ \text{NaOH} = \frac{0.10\ \text{mol NaOH}}{1\ \text{L solution}}$$

or

$$0.10\ M\ \text{NaOH} = \frac{0.10\ \text{mmol NaOH}}{1\ \text{mL solution}}$$

**Molar Concentration as a Critical Link Between Moles and Volume of a Solution**

The reason molarity is so useful for stoichiometry is because it relates the amount of solute in a solution to the volume of solution. This makes it easy to dispense moles of solute simply by measuring out volumes of solution.

*For calculations, molarity is the critical link that allows us to relate moles of solute and volumes of solution.* Molarity provides the conversion factors needed to calculate the number of moles of solute in a given volume of a solution, or the volume of solution that contains a certain number of moles.

Let's see how we can use molarity to form conversion factors. Suppose we have a solution that is labeled 0.250 $M$ HCl. The first thing we have to do is express this in units of moles and volume. For this solution we can write

$$0.250\ M\ \text{HCl} = \frac{0.250\ \text{mol HCl}}{1\ \text{L solution}}$$

For convenience, we'll write the relationship between moles and volume as an equivalence

$$0.250 \text{ mol HCl} \Leftrightarrow 1.00 \text{ L solution}$$

(Notice that we've expressed the volume to 3 significant figures. For molarity, the precision is indicated by the numerator; the denominator can be written to the same number of significant figures.) From this relationship we can form two conversion factors useful in calculations.

$$\frac{0.250 \text{ mol HCl}}{1.00 \text{ L solution}} \quad \text{and} \quad \frac{1.00 \text{ L solution}}{0.250 \text{ mol HCl}}$$

Let's look at two examples.

## PROBLEM 28
How many grams of KOH are in 50.0 mL of 0.168 $M$ KOH solution.

*Analysis:* We can express the problem in equation form as follows:

$$50.0 \text{ mL KOH solution} \Leftrightarrow ? \text{ g KOH}$$

Let's examine the data and see what we can calculate. We have both the molarity and volume of the solution. We know molarity relates moles and volume, so we should be able to use the molarity to calculate the number of moles of solute in the solution. We can write the molarity as

$$0.168 \, M \text{ KOH} = \frac{0.168 \text{ mol KOH}}{1 \text{ L solution}}$$

As written, this fraction is the conversion factor that will convert liters of solution to moles of KOH. However, to use it, we need the volume in liters, so we'll convert 50.0 mL to liters. In liters, the volume is 0.0500 L. Multiplying this volume by the molarity will give the number of moles of solute.

To convert moles of KOH to grams, the tool is the formula mass of KOH, which gives us the relationship,

$$1 \text{ mol KOH} = 56.10 \text{ g KOH}$$

Now we can see the path to the answer.

$$0.0500 \text{ L solution} \xrightarrow[\text{molarity}]{\boxed{\text{tool}}} \text{mol KOH} \xrightarrow[\text{mass}]{\boxed{\text{tool}}} \text{g KOH}$$

*Solution:* Let's apply the conversion factors

$$0.0500 \text{ L KOH} \times \frac{0.168 \text{ mol KOH}}{1 \text{ L solution}} \times \frac{56.10 \text{ g KOH}}{1 \text{ mol KOH}} = 0.471 \text{ g KOH}$$

The 50.0 mL of solution contains 0.471 g of KOH.

*Is the Answer Reasonable?* Let's do some proportional reasoning and approximate math. If we had an entire liter of the solution, it would contain 0.168 mol of KOH. One tenth (0.1) liter would contain 0.0168 mol, and half of that (0.05 L) would contain 0.0084 mol. This is a little less than 0.01 mol, so we'll use 0.01 as an approximate value. One mole of KOH weighs 56 g, so 0.01 mol would weigh 0.56 g. We have a little less than 0.01 mol, so the answer should be a little less than 0.56 g. The value we obtained (0.471 g) seems to be about right, so the answer is reasonable.

An important point to learn from the preceding problem is that whenever you have the molarity and volume of a solution, you can calculate the number of moles of solute. Just multiply the molarity by the volume (in liters).

$$\text{L solution} \times \frac{\text{mol solute}}{\text{L solution}} = \text{mol solute}$$

$$\underset{\text{volume (L)}}{\diagup} \qquad \underset{\text{molarity}}{\diagdown}$$

Now let's look at another problem.

### PROBLEM 29

How many milliliters of 0.135 $M$ $Na_3PO_4$ contains 12.5 g of the solute?

*Analysis:* The problem, in equation form, is

$$12.5 \text{ g } Na_3PO_4 \Leftrightarrow ? \text{ mL solution}$$

The molarity (0.135 $M$ or 0.135 mol $L^{-1}$) will allow us to relate volume of solution and moles of solute. The relationship is

$$0.135 \text{ mol } Na_3PO_4 \Leftrightarrow 1.00 \text{ L solution}$$

Before we can use it, however, we need to have the amount of solute expressed in moles. To convert grams to moles, we use the formula mass.

$$1 \text{ mol } Na_3PO_4 = 163.94 \text{ g } Na_3PO_4$$

We now have a way to go from g $Na_3PO_4$ to liters of solution. After we obtain this, we move the decimal point to change to milliliters.

*Solution:* We begin by applying the unit conversions.

$$12.5 \text{ g } Na_3PO_4 \times \frac{1 \text{ mol } Na_3PO_4}{163.94 \text{ g } Na_3PO_4} \times \frac{1.00 \text{ L solution}}{0.135 \text{ mol } Na_3PO_4} = 0.565 \text{ L solution}$$

Finally, we convert the volume to milliliters to obtain 565 mL.

*Is the Answer Reasonable?* One mole of $Na_3PO_4$ weighs about 160 g, so 0.1 mol would weigh about 16 g. Our amount (approximately 12 g) is about 3/4 of 16 g, or about 0.075 mol. If we round the molarity (0.135

*M*) to 0.15 *M*, then the amount of solution needed to have 0.075 mol would 0.5 L. Our solution is a little less concentrated than 0.15 *M*, so the volume required would be somewhat larger than 0.5 L. Our answer is "in the right ball park," so it's reasonable.

**Additional Problems**

1. A physiological saline solution contains 0.850 g of NaCl per 100 mL of solution. What is the molarity of NaCl in the solution?

2. What is the molarity of a solution that contains 50.0 parts per million (ppm) of $CaCO_3$? (For dilute solutions in water, 1 ppm = 1 mg/L).

3. Calculate the number of grams of KBr needed to prepare 250.0 mL of a 0.100 *M* KBr solution.

4. How many milliliters of 2.00 *M* NaCl solution can be prepared from 1.00 g of NaCl?

5. Fluoridated water contains about 1.0 μg $F^-$ per liter of water. What is the molarity of $F^-$ in fluoridated water?

**Answers**

1. 0.145 *M* NaCl, 2. $5.00 \times 10^{-4}$ *M* $CaCO_3$, 3. 2.98 g KBr, 4. 8.56 mL, 5. $5.3 \times 10^{-8}$ *M* $F^-$

## Relating Volumes and Concentrations of Reacting Solutions

**Skills Analysis**

1. How many liters of 0.1000 *M* HCl would be required to produce 10.00 moles of $H_2$ gas, in the following reaction?

$$Zn(s) + 2HCl(aq) = ZnCl_2(aq) + H_2(g)$$

2. How many grams of baking soda ($NaHCO_3$) are required to completely react with exactly 100 mL of vinegar (0.500 *M* $HC_2H_3O_2$ solution)? The reaction is

$$NaHCO_3 + HC_2H_3O_2 \rightarrow CO_2 + H_2O + NaC_2H_3O_2$$

3. What is the molarity of a solution of hydrochloric acid, if 25.00 mL of solution is required to completely react with 0.1000 g of $Na_2CO_3$? The reaction is

$$Na_2CO_3 + 2HCl \rightarrow 2NaCl + H_2O + CO_2$$

4. How many milliliters of 0.1000 *M* $AgNO_3$ would be required to completely react with 10.00 mL of 0.0500 *M* $CaBr_2$? The reaction is

$$2AgNO_3 + CaBr_2 \rightarrow 2AgBr(s) + Ca(NO_3)_2$$

**Answers**

Study this entire section if you missed any of these questions.
1. 200.0 L HCl solution, 2. 4.20 g $NaHCO_3$, 3. 0.07548 *M* HCl. 4. 10.00 mL $AgNO_3$ solution

In previous problems dealing with reaction stoichiometry, you saw that the amounts of reactants and products could be expressed in grams or moles. Molarity and volume provide a third alternative, as illustrated in the following problem.

### PROBLEM 30

How many grams of calcium carbonate will react with 75.0 mL of 0.250 $M$ HCl? The equation for the reaction is

$$CaCO_3(s) + 2HCl(aq) \rightarrow CaCl_2(aq) + CO_2(g) + H_2O$$

*Analysis:* The problem can be expresses as

$$75.0 \text{ mL HCl solution} \Leftrightarrow ? \text{ g CaCO}_3$$

We're dealing with a chemical reaction, so the critical link between HCl and $CaCO_3$ is provided by the coefficients of the equation. These give us the mole relationship between $CaCO_3$ and HCl.

$$1 \text{ mol CaCO}_3 \Leftrightarrow 2 \text{ mol HCl}$$

We have the volume and molarity of the HCl solution, which will allow us to calculate the number of moles of HCl. We just need to change 75.0 mL to liters.

We now know how we are going to go from the volume of the solution to moles of $CaCO_3$. To find grams of $CaCO_3$, the tool will be the formula mass $(100.1 \text{ g mol}^{-1})$

We can diagram the flow of the problem as follows:

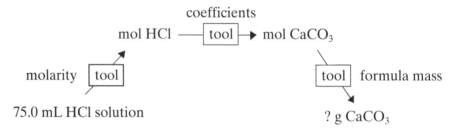

*Solution:* We'll start by changing the volume of the solution to liters: 75.0 mL = 0.0750 L. Then we apply the conversion factors following the path through the problem.

$$0.0750 \text{ mL solution} \times \frac{0.250 \text{ mol HCl}}{1.00 \text{ mL solution}} \times \frac{1 \text{ mol CaCO}_3}{2 \text{ mol HCl}}$$

$$\times \frac{100.1 \text{ g CaCO}_3}{1 \text{ mol CaCO}_3} = 0.938 \text{ g CaCO}_3$$

*Is the Answer Reasonable?* Let's suppose we had 100 mL (0.1 L) of the HCl solution. One liter of the HCl solution contains 0.25 mol HCl, so in 0.1 L, there would be 0.025 mol HCl. This would consume 0.0125 mol $CaCO_3$ which would weigh about 1.25 g. We have somewhat less than 100 mL, so we expect an answer somewhat less than 1.25 g. The answer we obtained, 0.938 g, seems to be reasonable.

We urge you to study Figure 3.11 on page 138 of the textbook. It maps out the paths of a variety of problems dealing with reaction stoichiometry, including those that involve molarity and solution volumes.

---

1. Find the concentration of $Fe^{2+}$ in a solution, if 25.00 mL of the solution reacted completely with 10.00 mL of 0.1000 $M$ $K_2Cr_2O_7$ solution. The reaction is

$$6Fe^{2+} + K_2Cr_2O_7 + 14H^+ \rightarrow 6Fe^{3+} + 2Cr^{3+} + 7H_2O$$

2. What is the concentration of silver in an $AgNO_3$ solution, if 25.00 mL of the solution reacts completely with 0.1000 g of NaCl? The reaction is

$$AgNO_3 + NaCl = AgCl + NaNO_3$$

3. When 100.0 mL of a solution that is contaminated with $Pb^{2+}$ ions is mixed with 25.00 mL of 0.1000 M HCl, the following reaction occurred,

$$Pb^{2+} + 2HCl = PbCl_2(s) + 2H^+$$

and 0.0100 g of $PbCl_2(s)$ was collected. If all of the lead in the original solution was converted into solid lead(II) chloride, what was the molarity of $Pb^{2+}$ in the original solution?

1. 0.2400 $M$ $Fe^{2+}$, 2. 0.06845 $M$ $AgNO_3$, 3. $3.60 \times 10^{-4}$ $M$ $Pb^{2+}$    **Answers**

# SUMMARY PROBLEMS

The following problems require you to bring together a variety of skills covered in this problem solving guide. To solve them you will have to determine the tools you need and how to put them together to find the solutions. If you are able to solve these problems, you are well on your way to success in chemistry. If you find you are still struggling, don't give up. Learning to solve problems requires persistence, so you should go back and review the topics we've covered. All the problems here can be solved using the tools we've discussed.

1. A 1.00 g sample of pure radium-226 undergoes $3.7 \times 10^{10}$ disintegrations per second. Roughly what percentage of radium-226 atoms in a 1.00 g sample disintegrate every second?

2. If gold costs \$9.28 per gram, how many atoms of gold can you buy for \$1.00?

3. Sulfur dioxide ($SO_2$) is a poisonous gas that is released into the atmosphere by volcanoes and by the burning of sulfur-containing coal. The World Health Organization recommends a concentration of no greater than 0.5 mg $SO_2$/ kg air over 24 hours for maximum exposure. The concentration is often reported in g/m$^3$ (that is, g $SO_2$/ m$^3$ air). Convert 0.5 mg $SO_2$/ kg air into g $SO_2$/ m$^3$ air, when the density of air is 1.18 g/L.

4. There are about 38 pounds of gold dissolved in a cubic mile of seawater. If a proposed extraction plant can extract all the gold from $1.00 \times 10^2$ gallons of seawater per second, how many grams of gold will it recover each day? (1 gal = 4 qt; 1 qt = 946 mL; 453.6 g = 1 lb; 2.54 cm = 1 in.; 12 in. = 1 ft; 5280 ft = 1 mi)

5. The Antarctic ice sheet contains an estimated 7.0 million cubic miles of ice. If the entire ice sheet melted, how many feet would the average global sea level rise? The density of ice is about 0.90 g/cm$^3$. The density of water is about 1.0 g/cm$^3$. The total area of the world's oceans is about 134 million square miles.

6. Heme, an iron-containing molecule embedded in the blood protein hemoglobin, contains one atom of iron per heme molecule. If heme is 9.06% Fe by mass, estimate the molecular mass of heme.

7. The following reaction is used to extract gold from pretreated gold ore:

$$2Au(CN)_2{}^-(aq) + Zn(s) \rightarrow 2Au(s) + Zn(CN)_4{}^{2-}(aq)$$

If 1.00 g of powdered Zn is added to 100.0 mL of 0.1000 $M$ $Au(CN)_2{}^-$, how many grams of solid gold will be obtained?

8. Mispickel (FeSAs) is a common arsenic-containing mineral. Baking mispickel in a furnace releases the arsenic. If a rock is 10.0% mispickel by mass, how many grams of arsenic will a 1.000 kg rock release when baked?

9. How many pounds of NaF would be required to fluoridate a reservoir containing 3.50 million gallons of water? The desired concentration of F$^-$ in fluoridated water is 1.00 μg/L.

10. What is the molarity of acetic acid ($HC_2H_3O_2$) in vinegar? (Vinegar is 3.0% acetic acid by weight. Assume that the density of vinegar is 1.00 g/mL.)

11. A student mixed 1.00 mL of 0.1000 $H_2SO_4$ with 1.00 mL of 0.0500 M $Ba(OH)_2$. The following reaction occurred: $H_2SO_4$ + $Ba(OH)_2$ → $BaSO_4(s)$ + $H_2O$. How many grams of $BaSO_4$ were formed?

12. The amount of arsenic in a sample can be determined by converting the arsenic to $HAsO_2$ and then reacting it with an $I_2$ solution of known concentration: $HAsO_2 + I_2 + 2H_2O \rightarrow H_3AsO_4 + 2H^+ + 2I^-$. Suppose a 500.0 mg sample was treated and the resulting $HAsO_2$ required 25.00 mL of 0.01000 M $I_2$ for complete reaction. What is the percentage of arsenic in the sample?

---

**Answers to Summary Problems**

1. The number of atoms in 1.00 g of radium-226 is $1.00 \text{ g Ra} \times \dfrac{1 \text{ mol Ra}}{226 \text{ g Ra}} \times \dfrac{6.02 \times 10^{23} \text{ atoms Ra}}{1 \text{ mol Ra}} = 2.66 \times 10^{21}$ atoms Ra, so the percentage of disintegrations every second is roughly $\dfrac{3.7 \times 10^{10} \text{ disintegrations}}{2.66 \times 10^{21} \text{ atoms Ra}} \times 100\% = 1.4 \times 10^{-9}\%$.

2. $\$1.00 \times \dfrac{1 \text{ g Au}}{\$9.28} \times \dfrac{1 \text{ mol Au}}{197.0 \text{ g Au}} \times \dfrac{6.02 \times 10^{23} \text{ atoms Au}}{1 \text{ mol Au}} = 3.29 \times 10^{20}$ atoms Au

3. $\dfrac{0.50 \text{ mg } SO_2}{1 \text{ kg air}} \times \dfrac{10^{-3} \text{ g } SO_2}{1 \text{ mg } SO_2} \times \dfrac{1 \text{ μg } SO_2}{10^{-6} \text{ g } SO_2} \times \dfrac{1 \text{ kg air}}{1000 \text{ g air}} \times \dfrac{1.18 \text{ g air}}{1 \text{ L air}} \times$

$\dfrac{1 \text{ L air}}{1000 \text{ cm}^3 \text{ air}} \times \dfrac{1^3 \text{ cm}^3 \text{ air}}{(10^{-2})^3 \text{ m}^3 \text{ air}} = 590 \text{ μg } SO_2/\text{m}^3 \text{ air}$

4. $\dfrac{1.00 \times 10^2 \text{ gal seawater}}{1 \text{ s}} \times \dfrac{4 \text{ qt seawater}}{1 \text{ gal seawater}} \times \dfrac{946 \text{ mL seawater}}{1 \text{ qt seawater}} \times$

$\dfrac{1 \text{ cm}^3 \text{ seawater}}{1 \text{ mL seawater}} \times \dfrac{1^3 \text{ in.}^3 \text{ seawater}}{2.54^3 \text{ cm}^3 \text{ seawater}} \times \dfrac{1^3 \text{ ft}^3 \text{ seawater}}{12^3 \text{ in.}^3 \text{ seawater}} \times \dfrac{1^3 \text{ mi}^3}{5280^3 \text{ ft}^3}$

$\times \dfrac{38 \text{ lb gold}}{1 \text{ mi}^3 \text{ seawater}} \times \dfrac{453.6 \text{ g gold}}{1 \text{ lb gold}} \times \dfrac{60 \text{ s}}{1 \text{ min}} \times \dfrac{60 \text{ min}}{1 \text{ hr}} \times \dfrac{24 \text{ hr}}{1 \text{ day}} =$

$\dfrac{0.14 \text{ g gold}}{\text{day}}$

That's about $1.30 worth of gold per day at current (2002) prices.

5. Begin by taking the inverse of the densities: For water, $1/d$ = 1.0 cm³/1.0 g = 1.0 cm³/g; for ice, $1/d$ = 1.0 cm³/0.90 g = 1.1 cm³/g. Since the denominators are the same (1 g) we can write: 1.1 cm³ ice ⇔ 1.0 cm³ water. The volume ratio is independent of units, so we can also write 1.1 mi³ ice ⇔ 1 mi³ water. Using this as a conversion factor allows us to calculate the volume of water produced by the melting of the ice cap. $7.0 \times 10^6 \text{ mi}^3 \text{ ice} \times \dfrac{1 \text{ mi}^3 \text{ water}}{1.1 \text{ mi}^3 \text{ ice}} = 6.4 \times 10^6 \text{ mi}^3$ water. If the global sea level rise were spread evenly over the world's oceans,

the sea level would rise by the volume divided by the area: ($6.4 \times 10^6$ mi$^3$/$134 \times 10^6$ mi$^2$) = 0.047 mi. Converting to feet, 0.047 mi $\times$ (5280 ft/1 mi) = 250 feet.

**6.** $\dfrac{1 \text{ g heme}}{0.0906 \text{ g Fe}} \times \dfrac{55.847 \text{ g Fe}}{1 \text{ mol Fe}} \times \dfrac{1 \text{ mol Fe}}{1 \text{ mol heme}} = 616 \text{ g heme / mol heme.}$

If there are 616 g in one mole, the molecular mass must be 616.

**7.** $1.00 \text{ g Zn} \times \dfrac{1 \text{ mol Zn}}{65.39 \text{ g Zn}} = 0.0153 \text{ mol Zn}$

$100.0 \text{ mL soln.} \times \dfrac{10^{-3} \text{ L soln.}}{1 \text{ mL soln.}} \times \dfrac{0.1000 \text{ mol Au(CN)}_2^-}{1 \text{ L solution}} = 0.01000 \text{ mol}$

Au(CN)$_2^-$. The zinc would require $0.0153 \text{ mol Zn} \times \dfrac{2 \text{ mol Au(CN)}_2^-}{1 \text{ mol Zn}}$

= 0.0306 mol Au(CN)$_2^-$, so Au(CN)$_2^-$ is the limiting reactant.

$0.01000 \text{ mol Au(CN)}_2^- \times \dfrac{1 \text{ mol Au}}{1 \text{ mol Au(CN)}_2^-} \times \dfrac{196.97 \text{ g Au}}{1 \text{ mol Au}}$

= 1.970 g Au

**8.** $1.000 \text{ kg rock} \times \dfrac{1000 \text{ g rock}}{1 \text{ kg rock}} \times \dfrac{0.100 \text{ g FeAsS}}{1 \text{ g rock}} \times \dfrac{1 \text{ mol FeAsS}}{162.84 \text{ g FeAsS}} \times$

$\dfrac{1 \text{ mol As}}{1 \text{ mol FeAsS}} \times \dfrac{74.922 \text{ g As}}{1 \text{ mol As}} = 46.0 \text{ g As}$

**9.** $3.50 \times 10^6 \text{ gal} \times \dfrac{4 \text{ qt}}{1 \text{ gal}} \times \dfrac{946 \text{ mL}}{1 \text{ qt}} \times \dfrac{10^{-3} \text{ L}}{1 \text{ mL}} \times \dfrac{1.00 \ \mu\text{g F}^-}{1 \text{ L}} \times \dfrac{10^{-6} \text{ g F}^-}{1 \ \mu\text{g F}^-} \times$

$\dfrac{1 \text{ mol F}^-}{19.00 \text{ g F}^-} \times \dfrac{1 \text{ mol NaF}}{1 \text{ mol F}^-} \times \dfrac{41.99 \text{ g NaF}}{1 \text{ mol NaF}} \times \dfrac{1 \text{ lb NaF}}{453.6 \text{ g NaF}} = 0.0645 \text{ lb NaF}$

**10.** $\dfrac{0.030 \text{ g HC}_2\text{H}_3\text{O}_2}{1 \text{ g vinegar}} \times \dfrac{1.00 \text{ g vinegar}}{1 \text{ mL vinegar}} \times \dfrac{1 \text{ mL vinegar}}{10^{-3} \text{ L vinegar}} \times$

$\dfrac{1 \text{ mol HC}_2\text{H}_3\text{O}_2}{60.052 \text{ g HC}_2\text{H}_3\text{O}_2} = 0.50 \ M \ \text{HC}_2\text{H}_3\text{O}_2$

**11.** $1.00 \text{ mL H}_2\text{SO}_4 \text{ soln.} \times \dfrac{10^{-3} \text{ L soln.}}{1 \text{ mL soln.}} \times \dfrac{0.1000 \text{ mol H}_2\text{SO}_4}{1 \text{ L H}_2\text{SO}_4 \text{ soln.}}$

$= 1.00 \times 10^{-4} \text{ mol H}_2\text{SO}_4$

$1.00 \text{ mL Ba(OH)}_2 \text{ soln.} \times \dfrac{10^{-3} \text{ L soln.}}{1 \text{ mL soln.}} \times \dfrac{0.0500 \text{ mol Ba(OH)}_2}{1 \text{ L Ba(OH)}_2 \text{ soln.}}$

$= 5.00 \times 10^{-5} \text{ mol Ba(OH)}_2.$

Ba(OH)$_2$ is the limiting reactant.

$5.00 \times 10^{-5} \text{ mol Ba(OH)}_2 \times \dfrac{1 \text{ mol BaSO}_4}{1 \text{ mol Ba(OH)}_2} \times \dfrac{233.39 \text{ g BaSO}_4}{1 \text{ mol BaSO}_4} =$

0.0117 g BaSO$_4$

**12.** $25.00 \text{ mL I}_2 \text{ soln.} \times \dfrac{10^{-3} \text{ L soln.}}{1 \text{ mL soln.}} \times \dfrac{0.01000 \text{ mol I}_2}{1 \text{ L I}_2 \text{ soln.}} \times \dfrac{1 \text{ mol HAsO}_2}{1 \text{ mol I}_2} \times$

$\dfrac{1 \text{ mol As}}{1 \text{ mol HAsO}_2} \times \dfrac{74.92 \text{ g As}}{1 \text{ mol As}} = 1.873 \times 10^{-2} \text{ g As.}$ The percentage of

As in the 0.5000 g sample is: $\dfrac{1.873 \times 10^{-2} \text{ g As}}{0.5000 \text{ g sample}} \times 100\% = 3.746\% \text{ As}$